iPhone 12 Pro and 12 Pro Max
Cameras

A Simple Guide to Shooting Professional photographs and Cinematic Videos on your iPhone 12 Pro and 12 Pro Max

KONRAD
CHRISTOPHER

Copyright

Konrad Christopher
ChurchGate Publishing House
USA | UK | Canada
© Churchgate Publishing House 2020

While the advice and information in this book are believed to be true and accurate at the date of publication, neither the authors nor the editors nor the publisher can accept any legal responsibility for any errors or omissions that may be made. The publisher makes no warranty, express or implied, with respect to the material contained herein.

Printed on acid-free paper.

Printed in the United States of America
© 2020 by Konrad Christopher

Table of Contents

CHAPTER ONE

Mastering the iPhone 12 Pro CAMERA

While it is true that a camera is not the only thing that makes a good photographer, it is important to reiterate that a good camera can improve the quality of your shots and even go as high as extending your ability to use various methods to improve quality of the photographs and videos – especially now that there are tons of technologies applicable to digital photography. The same thing is applicable to the iPhone camera. The iPhone camera is able to match the type of shots obtainable from most digital cameras, especially with the incorporation of third party apps and handy photography tools in the phone camera. The top-notch camera in your device will enable you to shoot professional and awe inspiring personal pictures without necessarily holding some heavy digital cameras. To obtain the maximum benefit out of your device's camera, it becomes expedient to master your device's camera capabilities and some of the various features that you can use to shoot quality pictures using the default camera app and some third party photography apps, which will be explored in the latter part of this guide.

Get Easy Access to Your Camera

Your device – iPhone – is portable and can fit inside your pockets and bags; this is one edge it has over some heavy digital cameras which probably have more hands or gears to stabilize them. Because of the portability, it becomes easy for you to take pictures anywhere and at any time; all that is required is to bring the device out of your bag or pocket and start taking amazing shots. Nonetheless, you still have an understanding of some quick means of accessing your device's camera app if you want to be able to record those beautiful moments as they are happening. The iPhone camera app can be accessed in these ways;

Using Siri: Say things like "Hey Siri, open the Camera app."

Swipe to Open: From the lock screen, swipe toward the left to launch the Camera app, and then hold on the camera icon. With this, you will be able to take pictures and use the camera app features even when your phone is locked. If your iPhone 12 is asleep, quickly press the device's side button to wake the device up. There is no sleep/wake button on the iPhone 12; hence, the side button is used to wake up the device.

Using the iPhone 12 Pro Control Center

Quickly swipe down from the top-right side of your device's screen to launch the iPhone Control Center. You will get some list of options where you can find your camera app and then tap on it to launch.

For quick access, you can even add the camera app to the Dock;

Add your camera application to the Dock

You will find the Dock just at the bottom of your device's main screen. Follow the steps below to add the camera app to the Dock;

- Go to the main screen of your iPhone where all of your apps are located.

- Tap and then lightly hold on to the camera app icon until the camera app starts to jiggle (shaking). The jiggling motion is preparing the app to be dragged to another location.
- Drag your camera app to the screen dock to be with the apps you most frequently used like your Safari browser and your Mail app. If the Dock is full already with apps, you can quickly remove an app that you don't use frequently from the Dock to accommodate the camera app.

The camera app will be placed beside the app highlighted above in the Dock for quick accessibility.

Multiple ways of taking pictures

There are about three ways of taking your shots with the iPhone 12 camera, each based on situations. The default method is by quickly tapping on the shutter button of your camera app to take the shot as shown below;

You can also use the iPhone volume up button or volume down button to take the picture as displayed in the image below;

With this, you can always hold your device properly to position it for better shots while taking pictures. Finally, if you have a wired headset, the volume buttons on the headset can be used to snap the shot. For example, you can capture good moments of friends and families at a party without them noticing to obtain some spontaneous pictures. You will be able to avoid shaking your device while taking your pictures when you used the headset buttons to snap your shot.

Enable the Grid Guides
It is best to know that the composition of any shots is part of what helps to determine how amazing the shot will be

in the end. Your iPhone camera app has grids that can help you while you are taking your shot to make sure the composition of the picture is naturally appealing. With the grid feature, you can divide your phone's screen into three different columns and rows. The Grid helps you to straighten your shot. Use the steps below to enable the grid feature on your device;

- Scroll to the **Settings app** 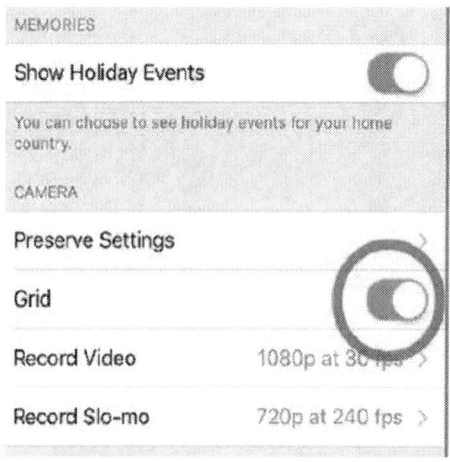 on your device.
- Scroll down a little bit and click on "**Photo & Camera.**"
- Toggle on the **Grid** feature in the camera section as shown below;

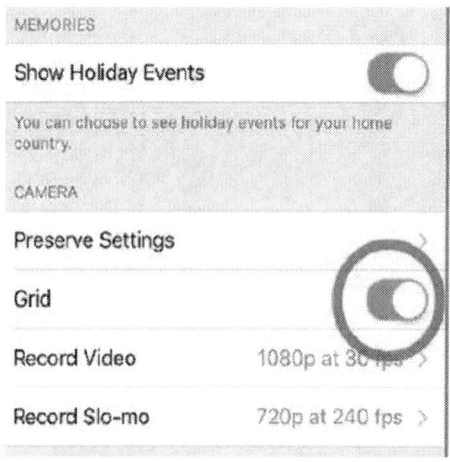

Shoot in Burst Mode

For events/actions that are happening fast such as sports or kids dancing at a party, you may wish to shoot multiple pictures of the event to enable you to pick the best picture later.

While tapping on your camera button repeatedly might be too slow, your iPhone's Burst mode can be used. With the burst mode, you can take multiple pictures almost immediately by tapping and holding on your phone's camera button. You can take more pictures in burst mode by holding on to the button much longer. Once you lift your hand off the button, the pictures will be saved in the Burst folder in the Camera roll. Follow the prompts below to access pictures in burst folder;

1. Navigate to your Photos app, and then click on the Burst folder to choose it.
2. Tap on "Select" in the middle-bottom toolbar to get the various pictures in the burst set.

To access one picture in the folder, do these;
1. Click on the picture and then select "**Done**" found at the top right part of your device's screen.

2. You can either click on "**Keep Everything**" to keep your folder and save the chosen image (or images) as a separate image or select "**Keep One.**"

How to adjust the flash on your device

1. From your device's home screen or Lock screen, launch the **Camera** app.

2. Tap on the **Flash toggle** located at the upper corner.

3. Choose whether you want the flash on **Auto**, **On**, or **Off**. There is an arrow at the top which you will tap to access **Auto**, **On**, or **Off** options.

Setting timer on your iPhone camera

1. From your device's Home Screen, open your camera app by tapping on its icon.

2. Click on the upper arrow found at the top part of your device's screen or simply swipe up from your camera's shutter button.

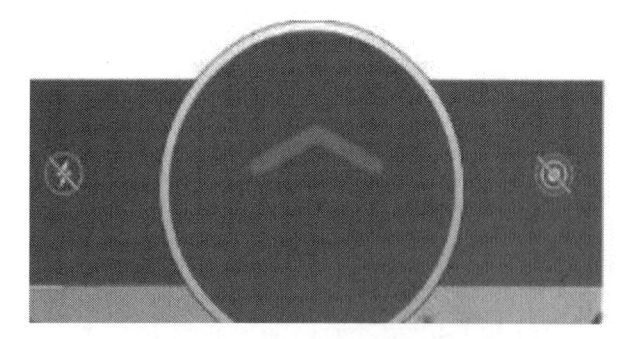

3. Click on the Camera's **Timer button**.

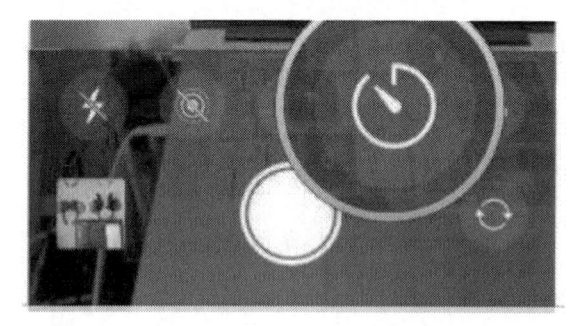

4. Choose between the **three seconds (3s)** or the **ten seconds (10s)**.

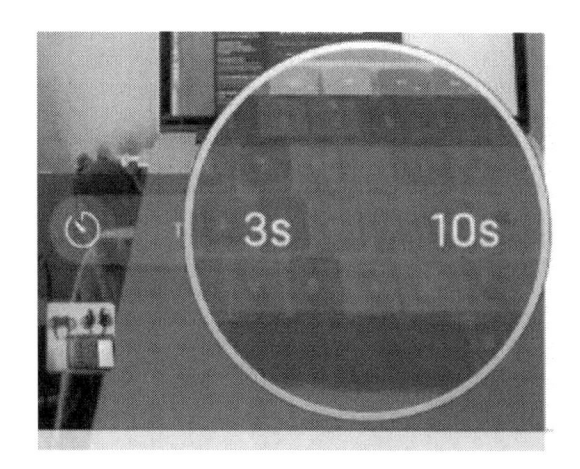

5. Begin the countdown by tapping on the **shutter icon** of your device's camera. You will see the screen starting to blink once the countdown has been initiated.

How to seamlessly switch from the rear-facing mode on

your camera to the front-facing option

1. From the Home screen, launch the camera app from your device's home screen by tapping on the icon of the camera.

2. Click on the **Flip camera icon** if you decide to move between the front-facing camera and the rear-facing camera.

3. Click on your camera's **Shutter button** to snap a picture or to start recording your video.

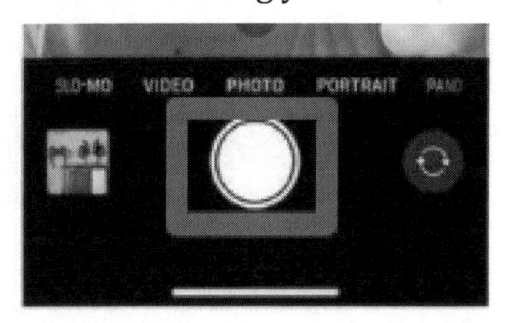

Set Up Manual Focus and Exposure

By default, your iPhone 12 camera will normally set the focus in your image toward the nearest object in the composition. Nonetheless, this setting can still be adjusted; for instance, you might want to focus your camera on a very far object in your shot. The iPhone 12 camera will give you the option to set the exposure and focus of your picture by tapping on the item that you want your camera to focus on and your camera will shift its focus to the object you selected. See the image below;

Adjust the camera's focus and exposure

Before you start snapping pictures on your iPhone, your device's camera automatically adjusts the focus and exposure, and the face detection will balance the exposure across many faces. The camera focus and exposure settings can be adjusted manually by carrying out the settings below;

1. Tap on your camera screen to bring the automatic focus area and the exposure setting.

2. Tap anywhere you plan to move the focus area.

3. Drag the ☀ up or down that is next to the focus area to adjust the camera exposure.

Lock the Exposure and Focus

Although your device's camera gives you the flexibility of setting the exposure and the focus in your shot, it still has a drawback, most especially when you are snapping a picture where the item may move. Or you might wish to snap multiple images of the same screen each time you take your shot without necessarily changing the focus.
To lock the manual focus and exposure settings for your upcoming shots, simply touch and then hold on the camera focus area until you get the **AE/AF lock;** then tap on your screen to unlock settings. You will be able to conveniently and accurately set and lock the exposure for your upcoming shots by clicking on the upper arrow icon 🔼 and then tap on ⊕ while moving the slider to adjust

exposure settings. The exposure will remain locked until you launch your camera. If you plan to keep your exposure control so that it won't be reset automatically on

launching the camera app, simply go to **Settings** , click on **Camera**, choose **"Preserve Settings,"** and then enable (toggle on) the **"Exposure Adjustment."**

Take low-light photos with Night mode

Use the Night mode feature on your device's camera to capture more details and then improve the brightness of your shot, especially in a low light condition. Although, in the Night mode, the length of the exposure is automatically determined, there is a way you can manage to adjust the length of the exposure.

On your device, you will find the Night mode on your device's front camera for selfie, on your Ultra Wide (0.5x) camera, and also on the Wide (1x) camera.

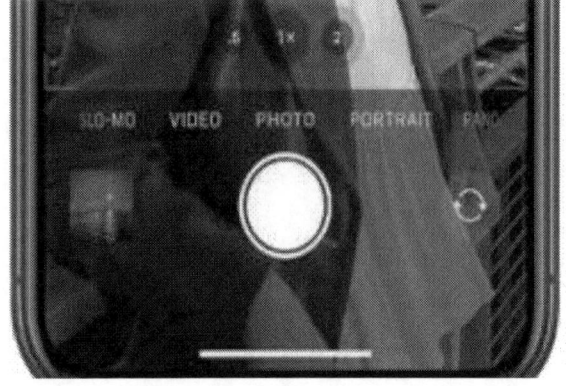

1. Choose the Photo mode from the available mode. If you are in a low-light situation, the Night mode will turn on automatically: you will observe the ⬤ button located at the top of your camera screen turning yellow and you will see a number next to the ⬤ button to show how many seconds the camera will spend to capture.

2. You can actually experiment with the Night mode feature by clicking on ⬤ and then utilize the slider right below the frame to select between the Max and Auto timers. With the Auto timers, the time will be automatically determined, while the Max will use the longest time. The setting you make here will be preserved for the next Night mode you are going to take.

3. Click on the camera's shutter icon and then hold the camera still to snap your image.

   There will be crosshairs inside the frame should the camera notice any movement during capture—you can actually align the crosshairs if you plan to reduce motions and improve the quality of your shot.

   Click on the **Stop button** just below the slider to stop capturing a Night mode shot midway.

Live Photo

It is rare to see the entire thing going on in most conventional pictures because some digital and phone cameras cannot really capture what is happening behind the scene. The Live Photo mode can be used on your device to capture fine details (what is happening in the background) just before you take the picture and right after the image must have been shot, including the sound in the background. In short, you will be able to take due cognizance of movements in your images with the live photo mode rather than freezing a moment with the still photo. The live photo will be able to record about 3 seconds of motion in your picture, for example about 1.5 seconds of movement is captured just before you tap on the camera's shutter button and about 1.5 seconds will be captured again once you click on the shutter button of your camera.

To take a live photo;

1. Make sure you are in the "Photo mode" among the list of camera modes at the lower end of the camera screen.

2. Click on the ◎ found at the top part (to the right) of the camera screen to enable/disable Live Photos.

3. Take the picture by tapping on your camera's shutter button.

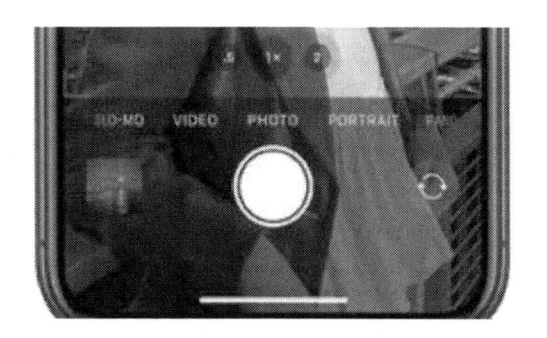

How to play a live photo

Once you have successfully snapped a live photo, the picture will be saved in your Photo app together with other still photos that you have snapped. But it is difficult to know whether an image is actually a live photo or not if you are viewing the list of images in thumbnail view. For instance, look at the image below and see if you can tell which of the images is actually a live photo and which one is not.

Obviously, it will be hard to know exactly which of the above pictures is actually a live photo and which one is not. One simple way you can use to know is actually by tapping on a picture to open such a picture in full screen. If the image is really a live photo, you will see the word **live** showing at the top part of your device's screen as shown below;

Editing a live photo

If you plan to edit a certain live photo, simply open the image and tap on the **edit** menu found at the top right side of the camera's screen and you will see the editing tools at the bottom of your screen;

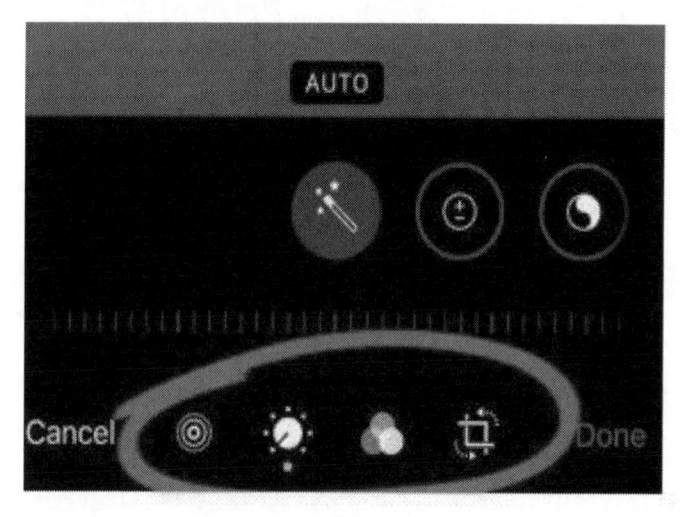

Choose the **adjust** icon if you wish to adjust the color, sharpness and exposure for your image by utilizing a range of tools that are available when you tap on the **adjust icon.**

You can then swipe from the right to the left and vice versa to select between different adjustment tools. Once you have selected the tool you want to use, you can then drag the slider to effect the adjustment.

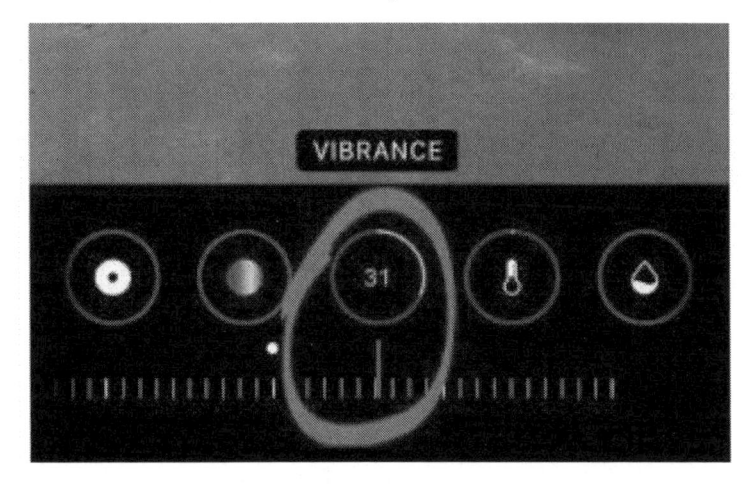

The **Filter** icon is used to apply color. To see the change, swipe between each filter and then apply the one you want.

You can crop or rotate the image with the **Crop** icon.

While cropping your picture, you can straighten the image by clicking on the **straighten** icon and then drag the slider either to the left or to the right. For perfect cropping of the image, there will be four corner crop handles for you to drag;

The icons located at the top left side of the screen can also be used to rotate/flip the picture;

How to use iPhone portrait mode to take amazing portrait photos

Sometimes, there are backgrounds in your final pictures which are not necessarily part of what you want to see in the after-image. The portrait mode is one shooting mode that you can use to blur out unnecessary backgrounds in your image.

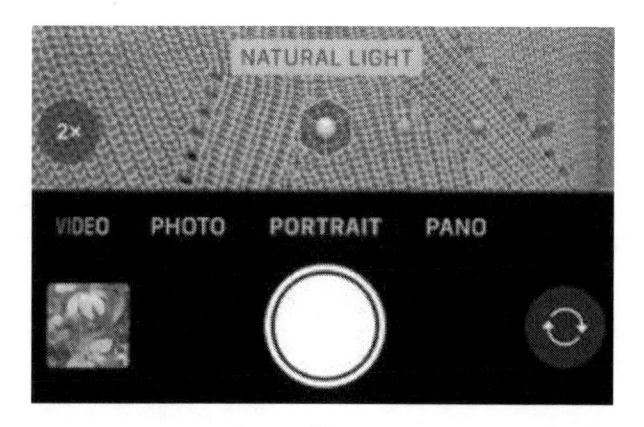

With the Portrait mode, the face of your subject (the person or subject you want to snap) will be very sharp and prominent while the background will be blurred.

With the background completely blurred, your subject will very much stand out while the look of the subject in the picture becomes something of emphasis. The portrait mode is useful – especially – when you have a very clumsy or not so attractive background. The background will still be there but it will be dreamy and the attention of the persons seeing the image will be totally drawn to the main subject in the picture.

To deploy the portrait mode, open your Camera app from the Home Screen or Lock Screen and then swipe through the various shooting modes at the bottom of the camera screen until you get the portrait mode in view.

While using the portrait mode, some important tips that you can consider include;

The portrait picture should be taken in places where there is adequate light. In case the light is not enough for the camera, you will be requested to get more light by moving to areas where there is adequate lighting.

In addition, your subject (the person or the item that you want to snap) must be between 2 feet to 8 feet away from the camera. If the object or the person that you want to snap is too close or even too far away, the camera will ask you to move far away or to draw closer. When you and the subject are at a decent distance away from each other, the word "Natural Light" will be shown in yellow. This means that your camera has been able to identify the subject and its background.

The camera can then automatically focus on the face of your subject and will then go ahead to blur out its background. Once the subject in the camera frame has been properly composed, simply tap on the shutter button of your camera to shoot the picture.

Once you have successfully snapped your photo, you can then go ahead to adjust the amount of blur in the background of the picture. You will be able to remove the blur completely or even add some studio effect.

How to change the level of blur with the Depth Control

The **Depth Control** functionality on your iPhone camera will allow you to adjust the level of the blur in the background of your picture just after you have taken the photo.

How to use the depth control

From the **photo app,** select a picture to view the picture in portrait mode, and then select **Edit** from the top right side of your screen.

Tap on the **f/number** icon found at the top left side of the screen.

The Depth slider will be displayed below your picture. Simply drag the Depth slider to the left or right to make your background blur stronger or weaker.

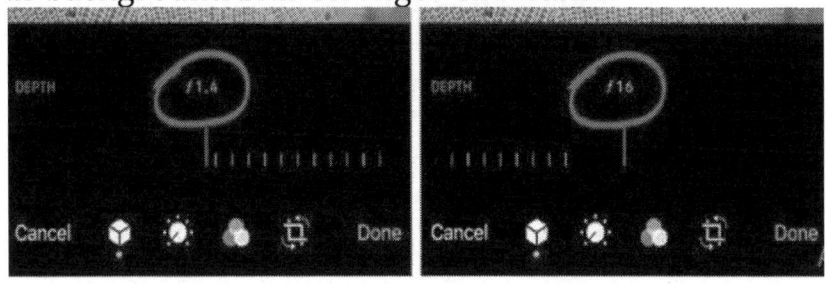

The smaller the f/number you chose, the blurrier the background of your photo appears. In the above illustration, the f/1.4 will give you a very blurry background, while the f/16 will bring more detail in the background. When you have successfully indicated the level of background you want to use, simply tap on "Done" to finish. The power of the blur can be modified anytime you want by using the Edit option once more.

Take a photo with a filter

Select the portrait mode or the Photo mode, click on ⌃, and then choose ⬤.

Below the viewer, swipe the filters to the right or left to preview the filter and then click on one of them to choose.

Shutter Speed: Changing Shutter speed on your device

Although, only DSLR cameras have inbuilt controls that can be used to control shutter speed, but you can use some third party apps on your iPhone to adjust shutter speed and produce excellent pictures.

What Is Shutter Speed?

Shutter speed implies the amount of time your camera's shutter is actually opened for taking a picture. The shutter speed is usually measured in seconds (or fractions of seconds), e.g. 8s, 2s, 1/30s, 1/250s, 1/500s, etc. A fast shutter speed implies that the shutter is actually opened for a very short amount of time (a fraction of a second).

Fast shutter speeds will freeze any motion in the scene. With a fast shutter speed, you can take good pictures of moving items or objects. Also, the amount of blurriness due to camera shake in your picture is reduced.

A slow shutter speed implies that the shutter is actually open for a much longer length of time (several seconds). If any item in the scene moves during this long exposure time, the movement will actually be recorded as a streak or blur – called motion blur.

Slow shutter speeds provide a perfect means of conveying a sense of motion in your picture. The motion blur will obviously indicate that your subject was actually moving through the frame.

A slow shutter speed will normally let you make a beautiful long exposure picture of rivers and waterfalls. The long exposure time makes the moving water come off as if it is silky smooth.

How to Change Shutter Speed On iPhone 12 Pro

The built-in Camera app that comes with your device does not contain a shutter speed option. Hence, there is a need to download a third party application that has manual controls compatible with iPhone.

There are many top-notch iPhone camera apps, which you can use to control the shutter speed.

The Camera+ 2 (sold at $2.99) is really a good option since it contains a decent range of shutter speeds you can choose from – from very fast to very slow. There is also a Slow Shutter shooting mode if you are planning to take long exposure pictures.

Check the process below to change the shutter speed using the Camera+ 2 app on your device:

Launch the **Camera+ 2** app and then click on the **[+]** icon just beside the shutter button to access shooting options. Make sure that the **Normal** shooting mode is chosen.

Tap on the **X** beside the shutter button if you want to close the shooting options.

Located just above the shutter button is the **ISO/Shutter Speed/** icon. It shows two values: the shutter speed (e.g. 1/33s) and ISO.

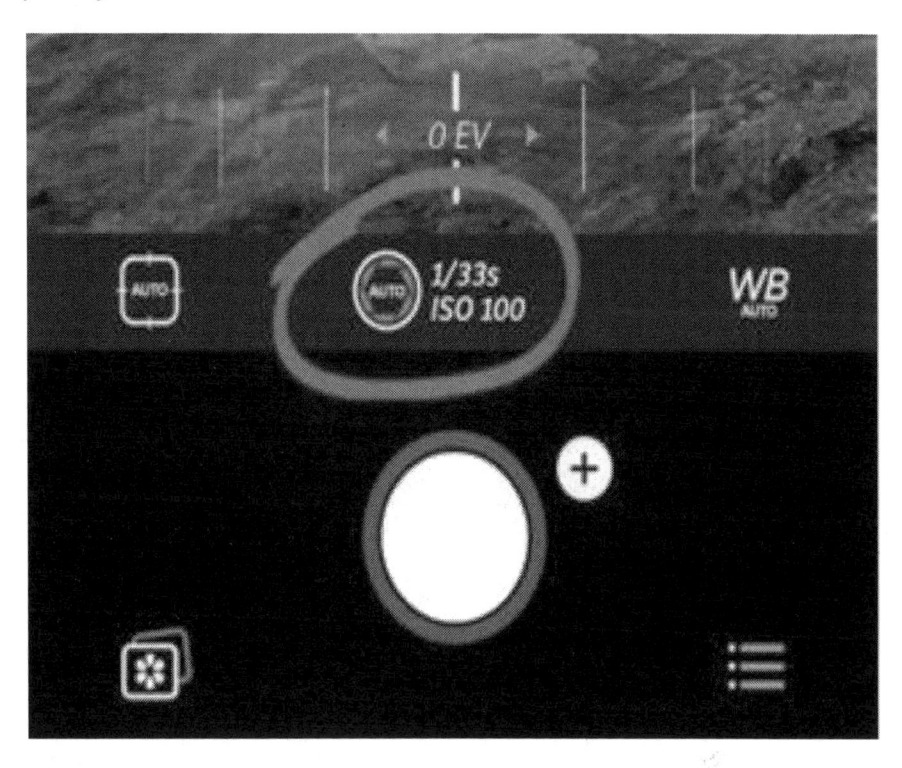

If you are unable to view the ISO values and the shutter speed, click on the **Menu** icon (three lines) just at the bottom right. Choose **Advanced Controls**, and then enable the "**Always Show**" by turning it on. Tap on the **Camera** icon just at the lower end of your screen to leave the menu.

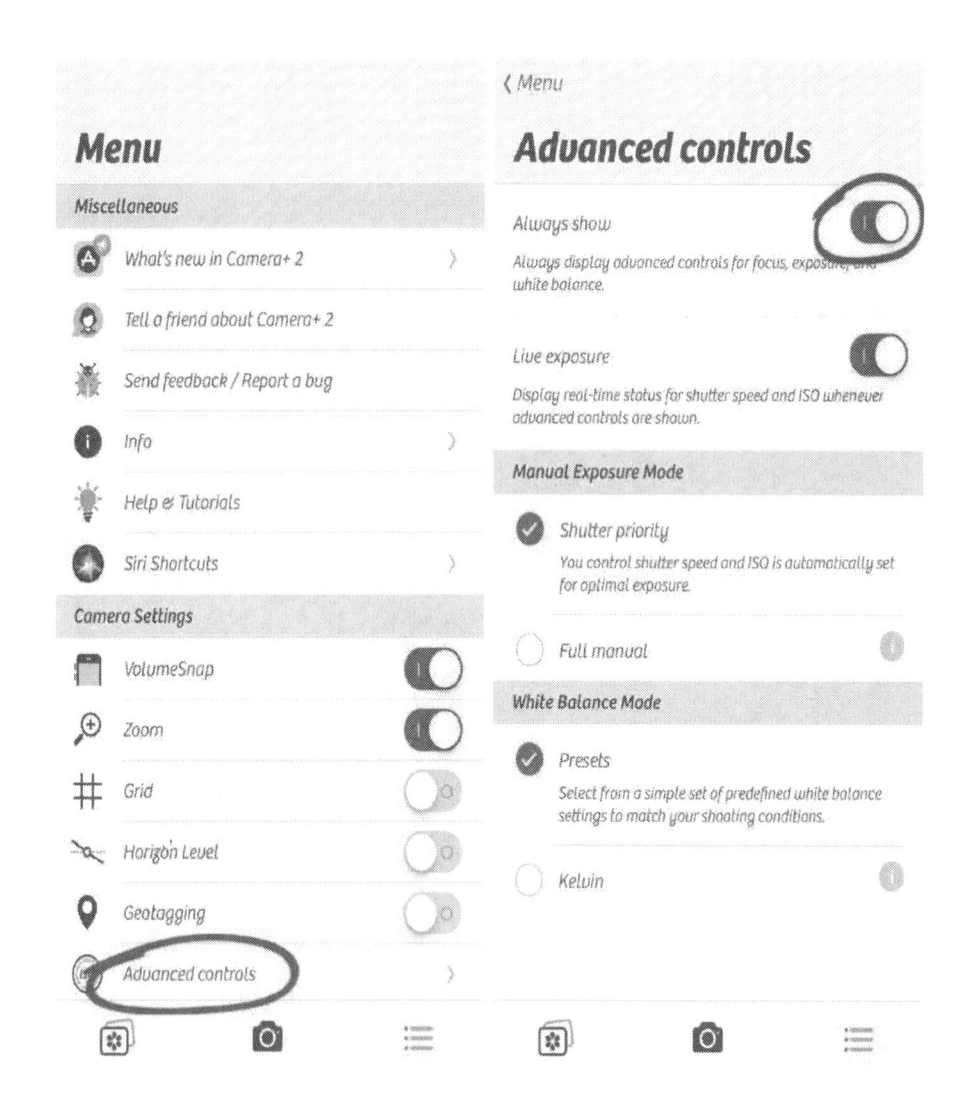

To change your shutter speed, simply click on the **Shutter Speed/ISO** icon just above the shutter button.

Doing this, you will get the **Shutter Speed** slider. You can adjust the shutter speed by dragging the shutter speed to the right or left.

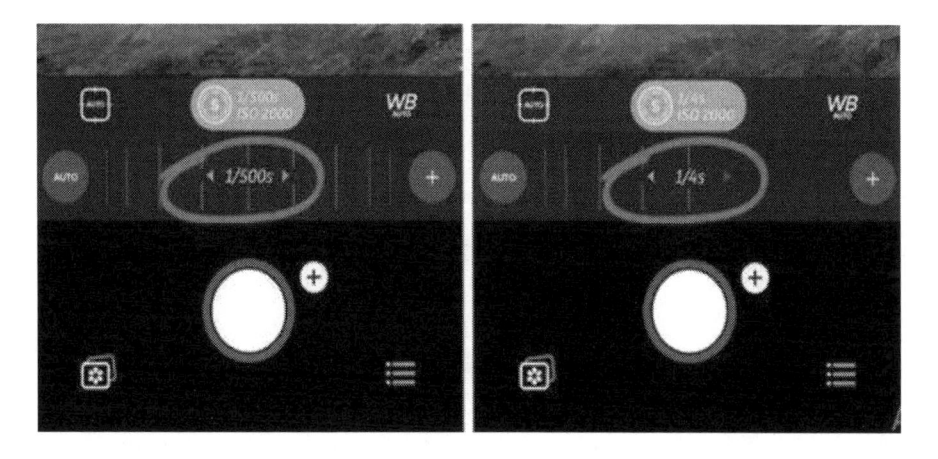

The further you go to the left, the faster your shutter speed tends. Faster shutter speeds will freeze the movement of moving items/subjects and will help you to eliminate/reduce camera shake.

The further you go to the right, the slower your shutter speed tends. A slower shutter speed will actually blur the movement of moving items or subjects.

You should understand that the Normal shooting mode has its slowest shutter speed set as 1/4s (a quarter of a second). If you plan to take long exposure pictures with the Camera+ 2, then you should use shutter speed slower than that.

Using a Slow Shutter Speed for Stunning Long Exposure Photos

In this part, you will see how you can capture amazing long exposure pictures by using two different and amazing apps: Camera+ 2 (sold at $2.99) and the Slow

Shutter Cam (sold at $1.99). Both of these apps yield excellent results, although they have slightly different functionality. Before anything, you will have to make sure that you have a tripod stand for your iPhone. The tripod stand will keep the camera still while you take your picture.

If you hold your device with just your hand, it won't be kept still enough and what you get will be a shaky picture that is totally blurred.

How to Shoot Long Exposure Photos Using Camera+ 2

The **Camera+ 2** app features a Slow Shutter mode for taking long exposure photos.

Tap on the **[+]** icon beside the shutter button, and then choose **Slow Shutter** mode.

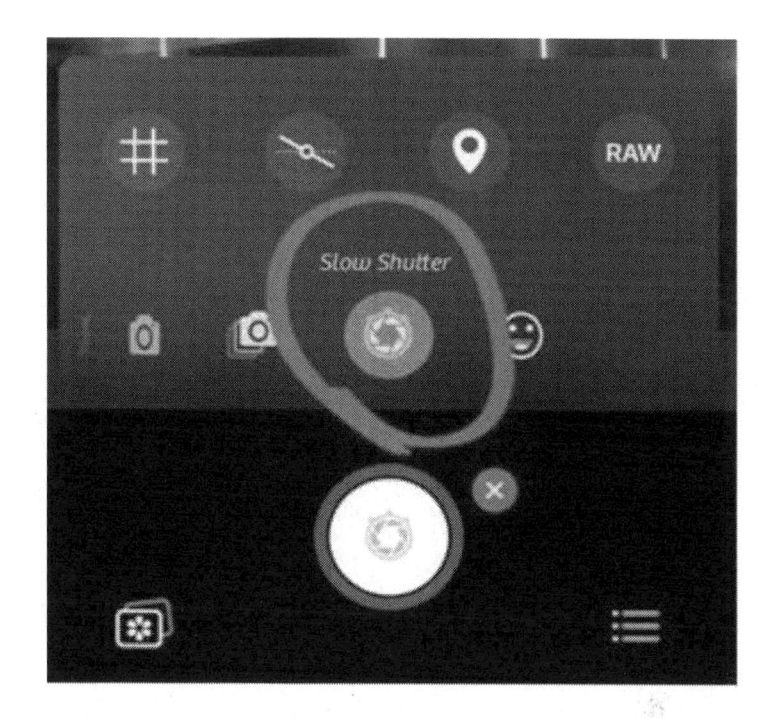

Click on the **X** that is beside the shutter button if you want to close the shooting options.

There are two sliders just above the shutter button. The slider on the left is the one that controls the shutter speed.

Set the value of the shutter speed by dragging the **Shutter Speed** slider. You can choose between 30 seconds, 15 seconds, 2 seconds etc. The longer your shutter speed, the more blurred any motion will look.

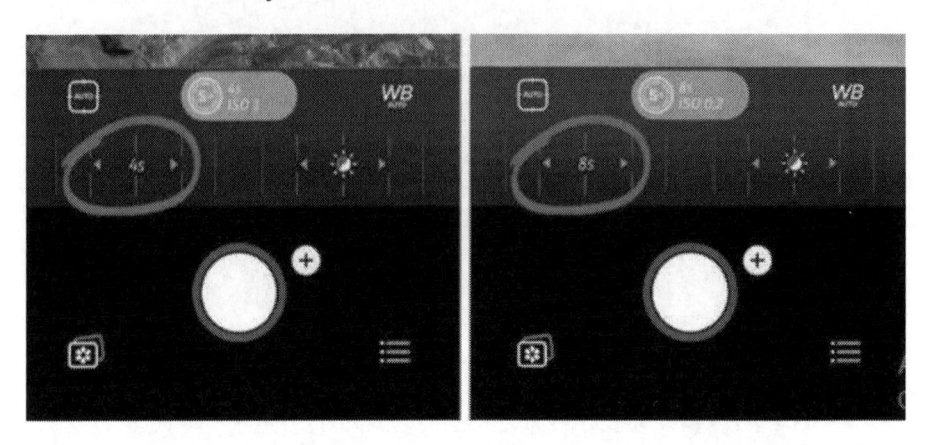

If you wish to adjust brightness (exposure), deploy the Sun icon right hand slider.

When you click on the shutter button of the app to capture your shot, you will need to make sure that the camera is still while you are snapping the picture.

How To Shoot Long Exposure images with the Slow Shutter Cam

The **Slow Shutter Cam** app is specifically made for creating long exposure photos.

With the app, you can take long exposure images of low light scenes, water and light trail.

Launch your **Slow Shutter Cam** app from the Home Screen. Tap on the **Settings** icon located at the lower left to access the capture mode options.

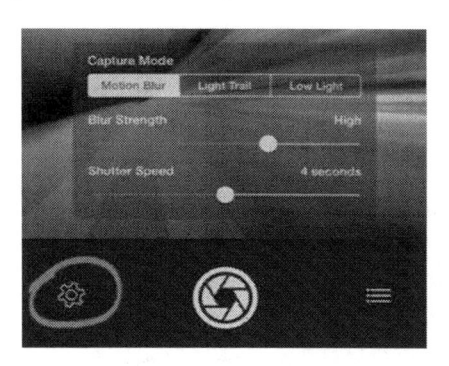

The app contains three essential slow shutter capture modes: **Motion Blur**, **Light Trail**, and **Low Light**.

The **Motion Blur** option is okay for making long exposure photos of running water, although it can still deployed to blur out the movement of other moving items or subjects.

The **Light Trail** mode will usually allow you to capture very beautiful long exposure images of moving lights. It is especially best for capturing car light trails and fireworks.

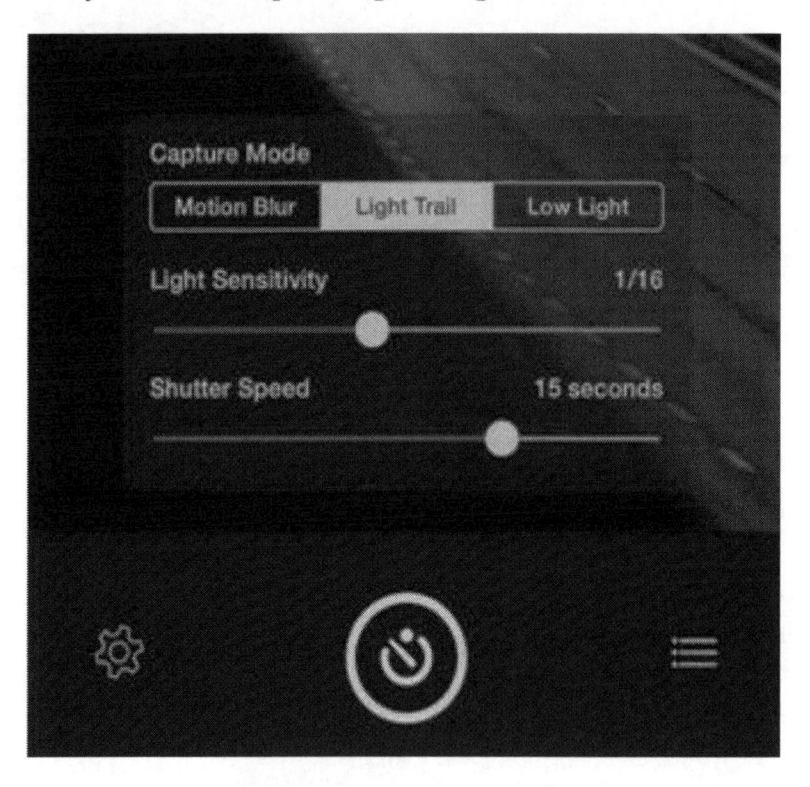

Sometimes, it might be important you take pictures at night; hence, the **Low Light** mode becomes handy for having brighter exposures when you are shooting photos at night.

Each of the capture mode selected contains a **Shutter Speed** slider. Use the Shutter speed slider by dragging it to the left to have a very fast shutter speed or achieve a slower shutter speed by dragging it to your right.

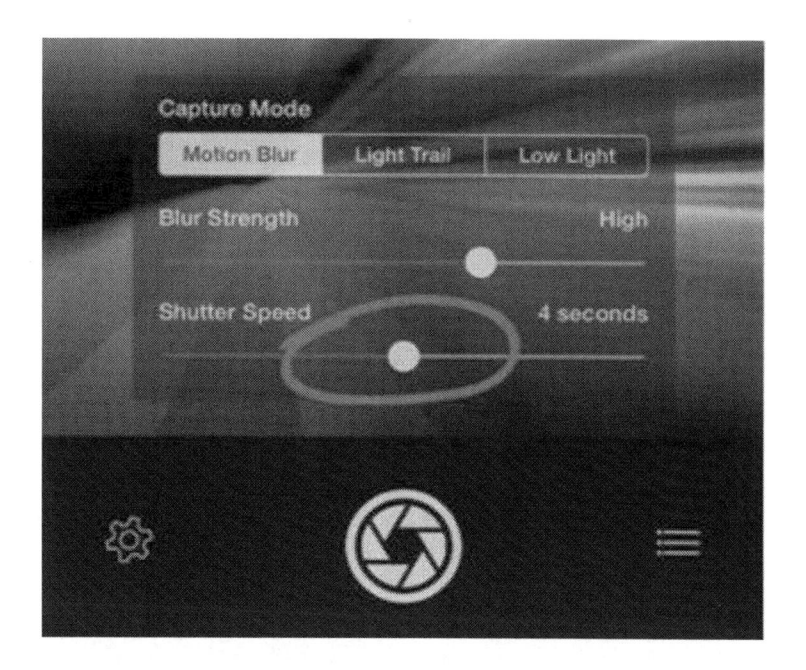

To enable the Bulb mode, simply drag the slider of the **Shutter Speed** towards the right. With this, you can take your pictures using any shutter speed of your choice. Click on the Shutter button at the bottom to initiate the exposure, and click on the shutter button once more to end taking your photo.

You can further fine-tune your photo by using the other sliders. For instance, you will be able to have control over how the motion appears with the **Blur Strength** slider. The highest motion is achieved when you drag the **Blur strength** slider to the extreme right.

Once you have successfully snapped your photo, select "**Edit**" from the lower end of your screen to further edit the image. Use some of the icons displayed at the bottom end of your screen to carry out further editing of the picture, then select "**Done**."

Select **Save** if you wish to save the photo. Or just click on "**Clear**" to discard the picture.

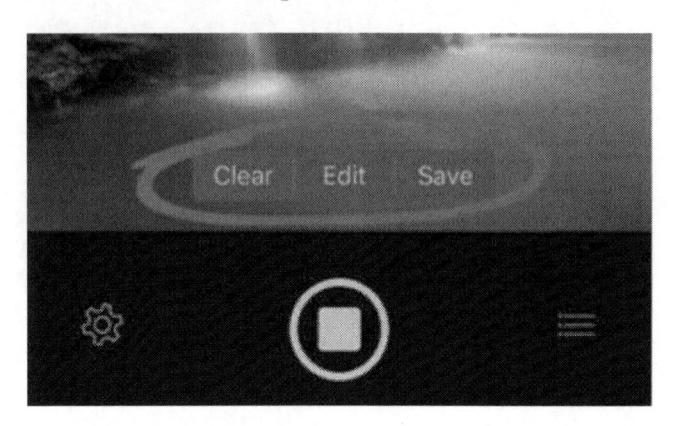

You can only imagine the level of creativity that can be achieved with your iPhone camera by deploying a slow shutter speed on your device.

ISO: How to Use ISO to Capture Grain-Free Shots in Low Light

ISO is what controls how sensitive your camera's sensor will be to light. The higher the value of the ISO, the more sensitive the sensor of your camera will be to light. This can be useful, especially, when you are shooting in a low light situation. However, using a high ISO has some disadvantages. Unfortunately, a high ISO setting will bring unwanted grain in your image (as shown below). The grain will be more pronounced in dark sections of the picture;

Grain lowers the quality of your picture. So it is better you avoid high ISO settings as much as possible.

How to Change the ISO iPhone Camera Setting

The built-in Camera app that came with your iPhone does not have the capability to allow you to change ISO. Instead, it will automatically choose an appropriate ISO setting, which depends on the amount of light in the scene.

In the low light situation, your camera will mostly set a high ISO in order to take more lights. This is why, sometimes, you get a grainy picture when you take pictures in dark mode.

However, there is a method that can be used to have control of your ISO iPhone settings; and once do this; you can always select a lower ISO in order to avoid having grainy pictures.

Check the process below to adjust your camera's shutter speed with the Camera+ 2 app on your device:

Launch the **Camera+ 2** app and then click on the positive **[+]** icon just beside the camera's shutter button to access the shooting options. Make sure that the **Normal** shooting mode is chosen.

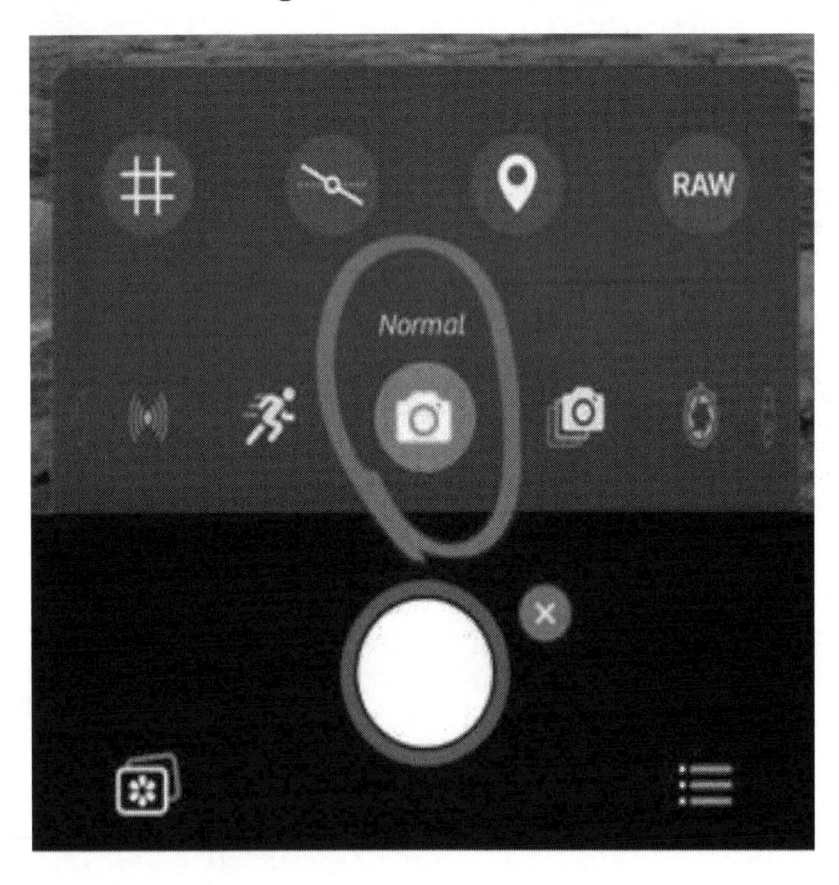

Select the **X** sign just beside the shutter icon if you plan to terminate the app's shooting options.

Located just above the camera's shutter button is the **ISO/Shutter Speed/** icon. It shows two values: the current value of the camera's shutter speed (e.g. 1/33s) and ISO.

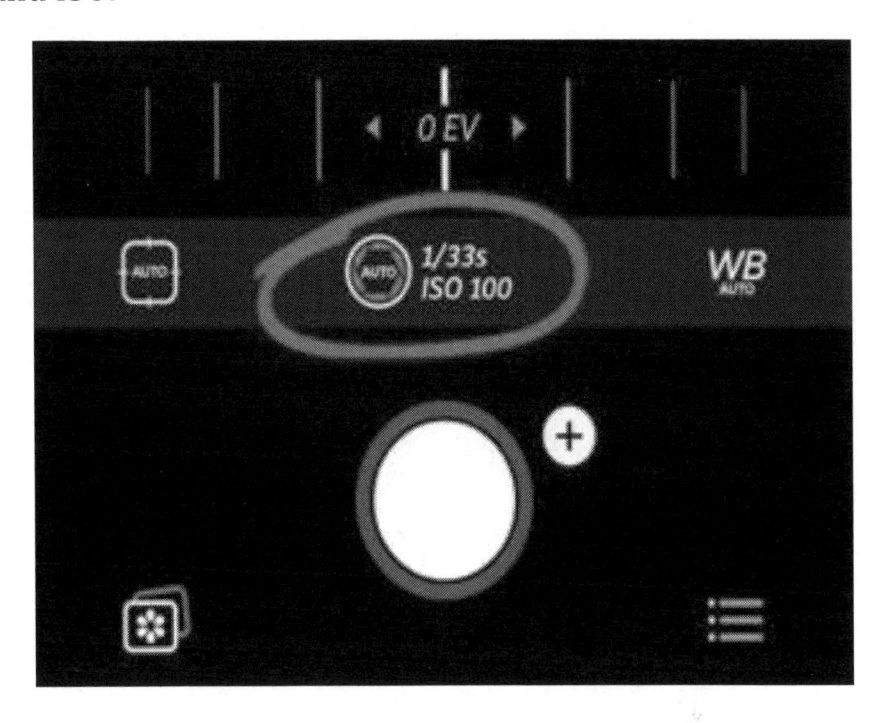

If you are unable to access the ISO values and the shutter speed, you can always click on the **Menu** icon (which is shown in three lines), choose **Advanced Controls**, and then enable the "**Always Show**" by turning it on. You can leave the menu by tapping on your camera icon just at the lower end of your screen.

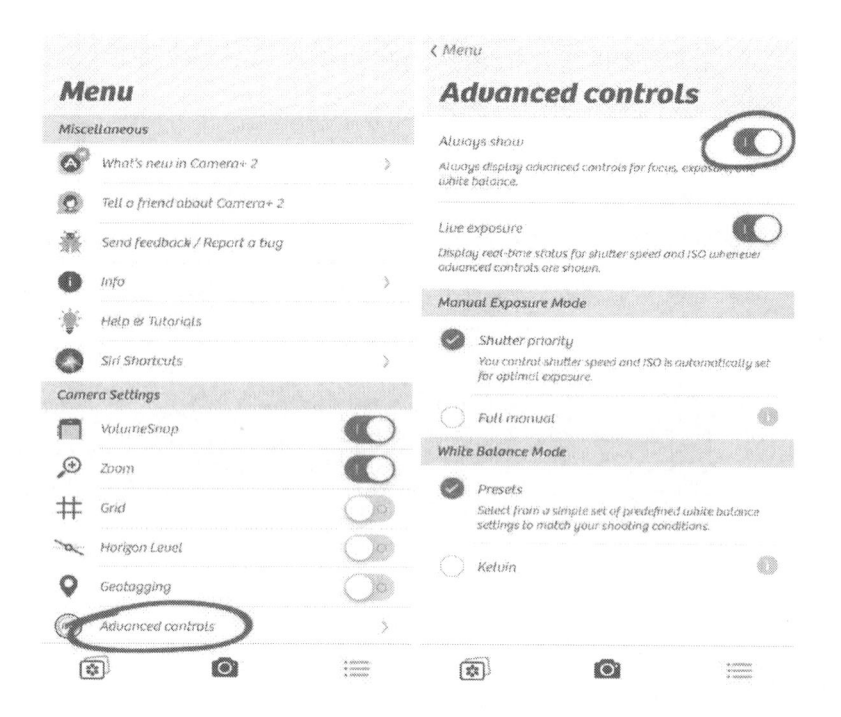

If you plan to adjust/change the ISO, simply click on the **Shutter Speed/ISO** icon just above your shutter button.

To begin with, it is only the slider of the Shutter Speed that is shown. To bring the adjustment slider for the ISO, simply click on the positive **[+]** icon just at the right of the adjustment slider.

You will get two sliders: which are the Shutter Speed slider on your left, and the ISO on your right. Simply drag the slider for the **ISO** to the left or to your right if you wish to adjust the ISO setting.

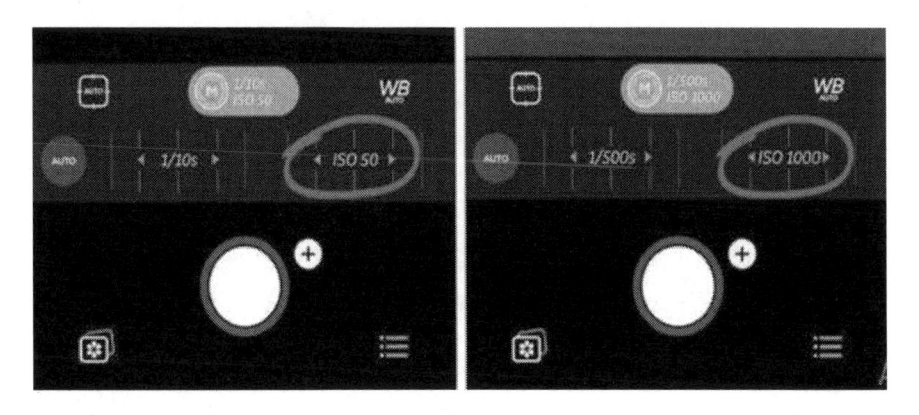

How to deploy a Low ISO to prevent Grainy Pictures

The higher value you set for the ISO, the more the light that your camera will be able to absorb, and the grainier your shot will actually appear. So it is essential you go for the lowest value of ISO settings that will get you a good exposure.

But things are not this straightforward, this is because once you lower the value of the ISO to a somewhat low value, the image will look dark.

However, you will be able to have your device's camera capturing more light by using a slower shutter speed. To do this, utilize the Shutter Speed slider at the left side.

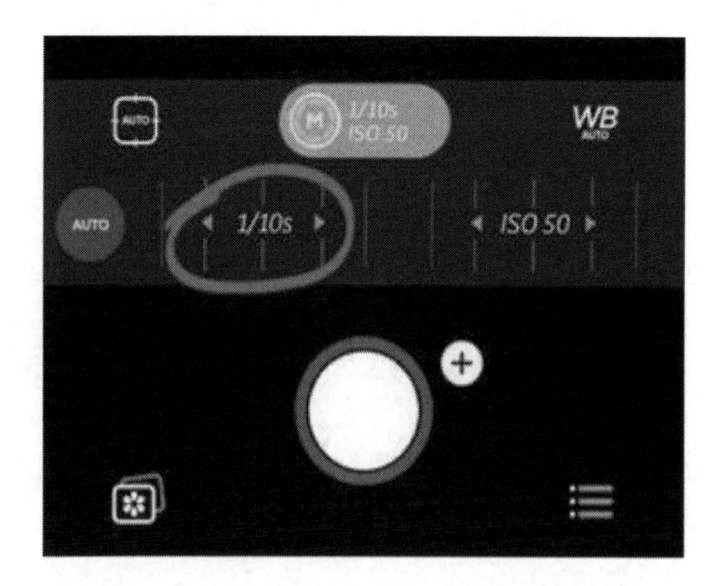

The slower shutter speed (which captures more light) will actually balance the low value of ISO (which captures less light). This enables you to deploy a low ISO value while still enjoying a decent exposure with adequate light.

White Balance: How to Capture Perfect Colors In Your Photos

The iPhone camera can only try it best by capturing colors, but most times, it is not that accurate as the colors might not be correct.

White balance enables you to record colors in your shots more accurately. It makes sure that the white sections of your image appear white, which eventually enables others colors in the picture to be recorded correctly.

Here are some cases where you are likely to get a strange color in your image;

If you snap pictures indoors while the light is on, you might sometimes observe an orange or a yellow color cast in your image. This is most often the result of the warm colored light coming from your light bulbs. Color casts become particularly evident in the white part of a picture.

If you snap your pictures outdoors, under cloud covers or in the shade, your pictures might come out having blue color cast. This is most times the case when you are taking pictures of snow.

When you are able to adequately adjust the white balance in your image, you will be able to completely eliminate or reduce any unnecessary color cast in your shots by warming up the colors or cooling down the colors in the image.

Choosing the exact white balance setting has a way of making sure that the white appears as white in your photo and other colors remain properly separated.

How you can change/adjust the White Balance on your iPhone 12 pro

The Camera feature that came with your device doesn't have the capability to allow you to change settings for the white balance. It can, instead, adjust the camera's white balance while trying to eliminate any cool or warm color cast in your picture.

The auto white balance option of your camera does a better job most times, especially when you are outdoors in natural light. But if you are snapping your pictures in a very cool or warm light, it then becomes necessary to have manual control of the White balance for proper color adjustment.

The white balance **preset** can be adjusted by using a third party app on your device called **Camera+ 2** app.

Launch your **Camera+ 2** app. Click on the **WB** (White Balance) icon located at the bottom right.

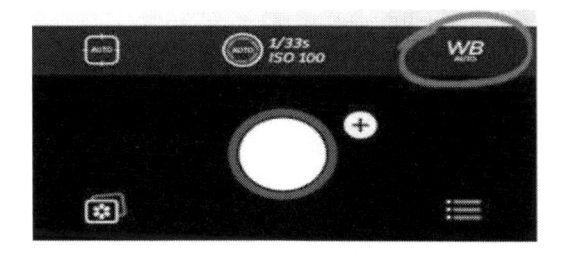

If you are unable to access the WB icon, click on the **Menu** icon (three lines), choose **Advanced Controls**, and then enable the "**Always Show**" by turning it on. Tap on the **Camera** icon located at the lower end of your screen to leave the menu.

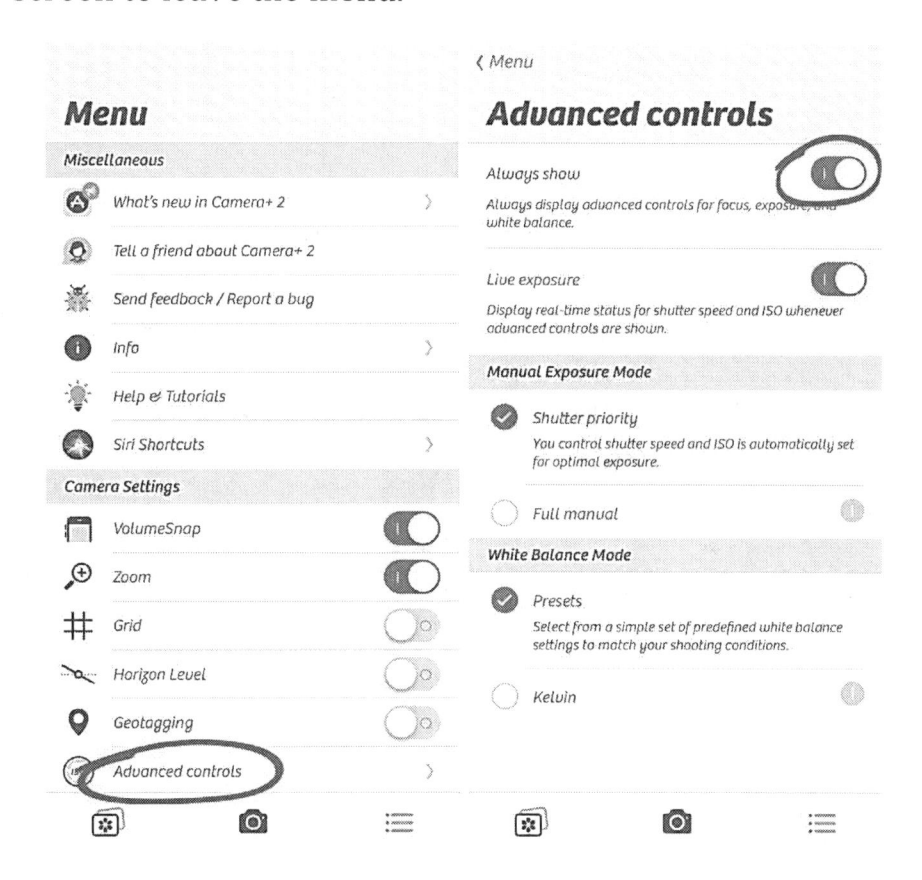

When you click on the **WB** icon, you will see a row of white balance **presets** coming up. The **presets** include; **Cloudy, Shade, Daylight, Flash** etc. You can access many white balance **presets** by swiping across.

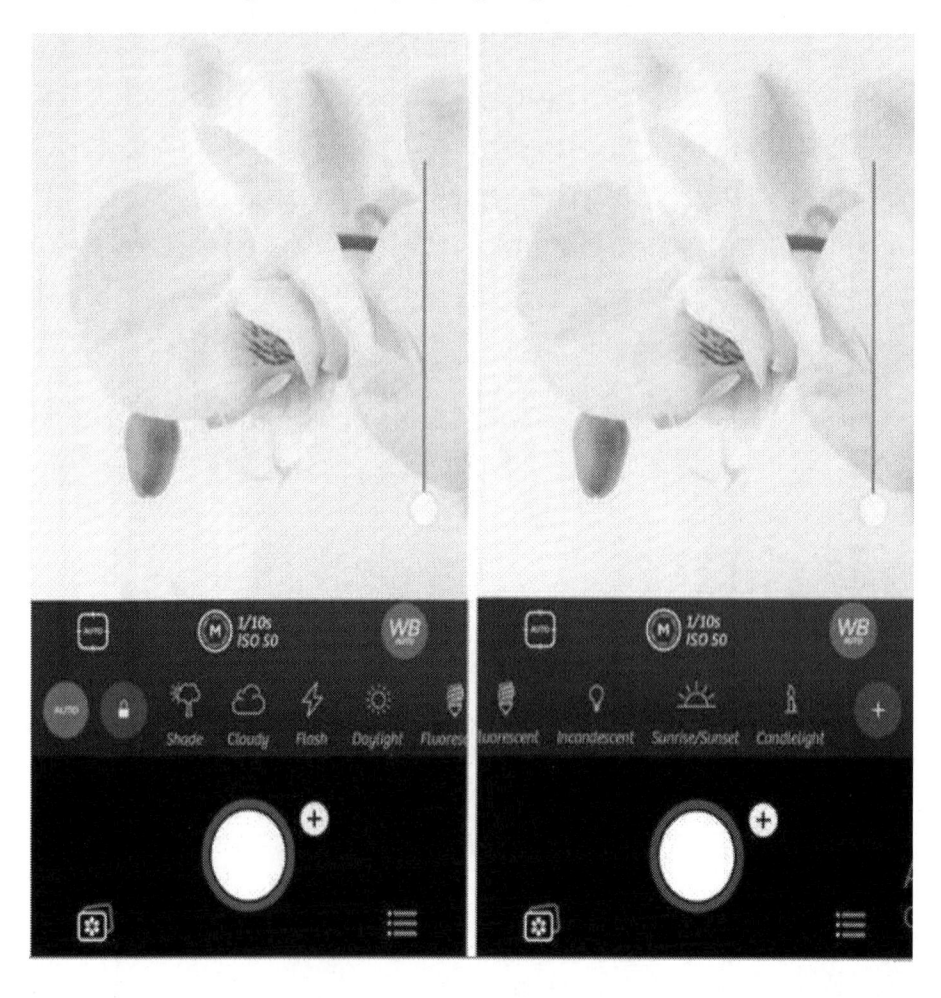

To remove the color cast in the flower image above, choose the white balance **preset** that exactly matches the kind of light you are shooting under.

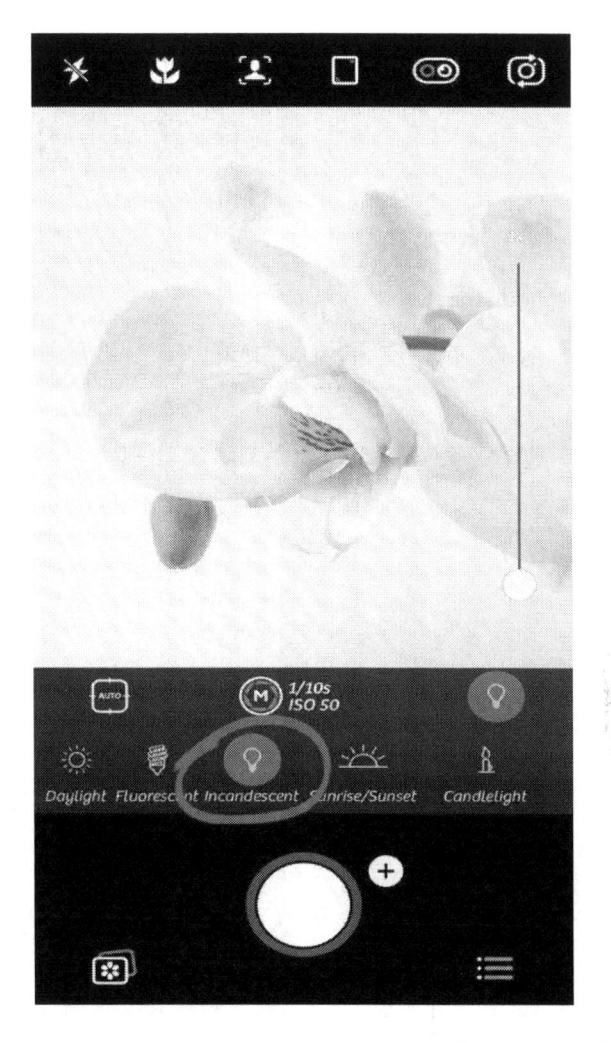

When you are taking your pictures under a warm-colored light bulb, select **Incandescent**, but when you are taking pictures in shade, choose the **Shade** white balance option.

Any warm or cool color cast from the light source will be eliminated when the camera adjusts the colors. In the case below, the **Shade** white balance preset was chosen. This setting will warm up the colors, and help to eliminate any blue color cast in your picture.

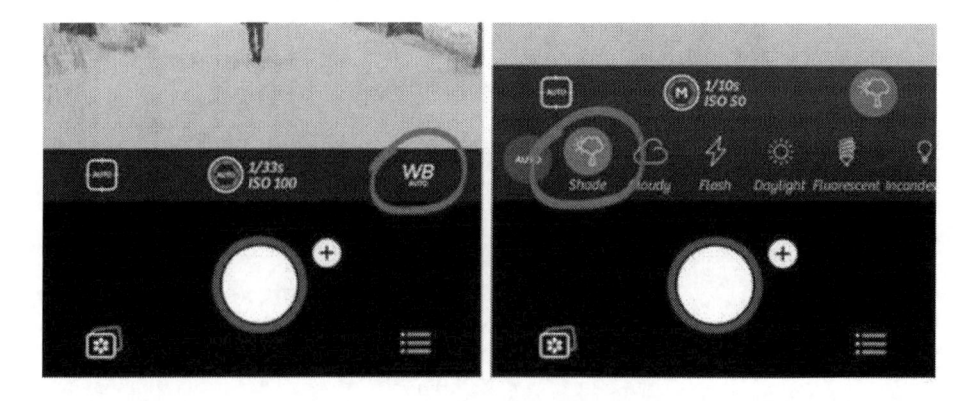

In the example below, the **Incandescent** preset was chosen. This setting actually cools down the colors, and then helps to neutralize the warm color cast from the light bulbs.

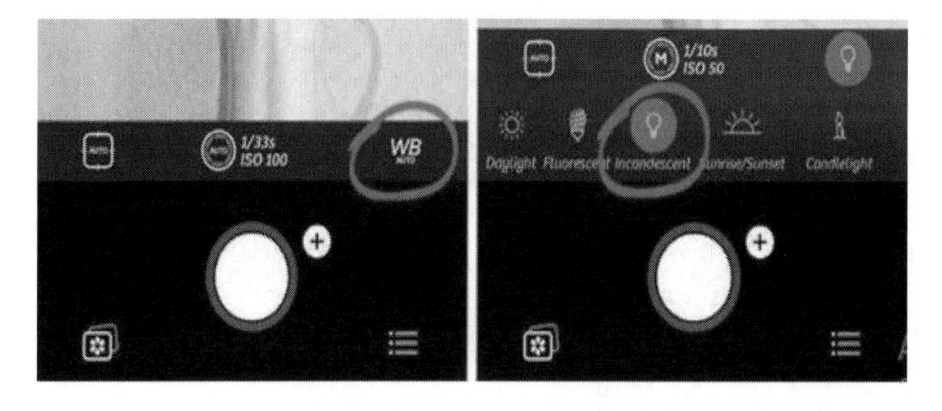

There are times when none of the presets will deliver a decent result, if this is the case, manually adjusting the color balance is the right step.

Click on the **[+]** icon located to the right side of the white balance presets. Then utilize the slider for the **White Balance** to cool down or to warm up the photo.

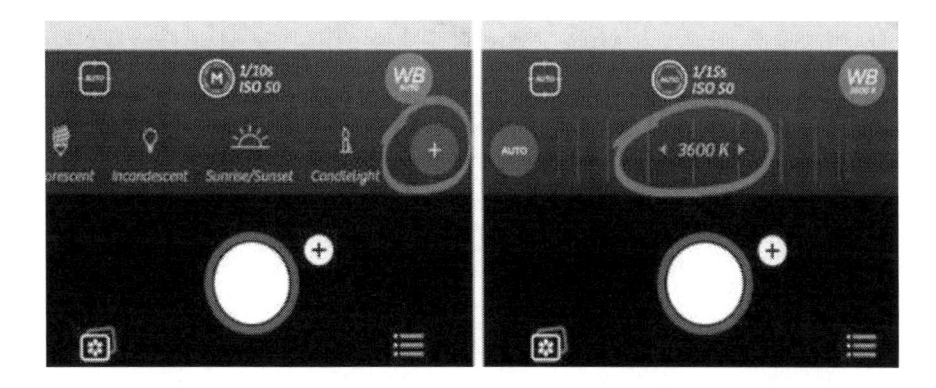

What you are aiming for by experimenting, as this is to get a white balance setting in which the white will appear as white in your picture.

If you want to have your camera deciding on which white balance setting to deploy, click on **Auto** situated to the left side of your white balance presets.

Use White Balance for Creative Effect

Getting all colors in your pictures to appear as accurate as possible is one thing you should consider to have uniform tone and quality pictures. For instance, you will always want the white to appear just as white. But sometimes, you can enhance a picture by cooling down or warming up the colors. For example, you will be able to emphasize the bleakness and coldness of a winter landscape when you cool down its colors.

Or you can even add more intensity to the color of sunset when you add warm tones to the photo.

When you understand how to use White balance accurately, you will be able to incorporate different moods in your shot.

CHAPTER TWO

Take videos with your iPhone 12 camera

While it is easy to snap amazing pictures with the iPhone camera, it becomes especially interesting to know that shooting videos with your iPhone camera is much easier. All that it takes is getting it right with the iPhone camera app .

Record a video

1. Swipe through the list of shooting modes available at the lower end of your camera screen and choose Video mode.

2. Click on the Record button or you can alternatively use either the volume up or volume down button to begin recording. You will still be able to carry out these actions while shooting a video;

   - Snap a still picture by clicking on the white Shutter button.

   - Pinch the screen to zoom in and out on the screen.

   - To arrive at a more defined zoom, simply touch and then hold the 1x while adjusting the slider to the left.

To terminate recording, simply click on the camera record button or you can alternatively use either the volume down or volume up button.

Videos shot with your iPhone camera normally record at 30 frame rate per seconds (fps). You can always choose from a different frame rate by going to **Settings,** tap on **Camera** and click on "**Record Video.**"

Use quick toggles to change video resolution and frame rate

In the Video mode, the quick toggle can be used to change the video resolution and the video frame rate. The quick toggle is found at the top area of the screen.

On the camera app, tap on the quick toggles located at the top-right side to select between the 4K or HD mode and 60, 30, or 24 frames per second (fps) in the Video mode.

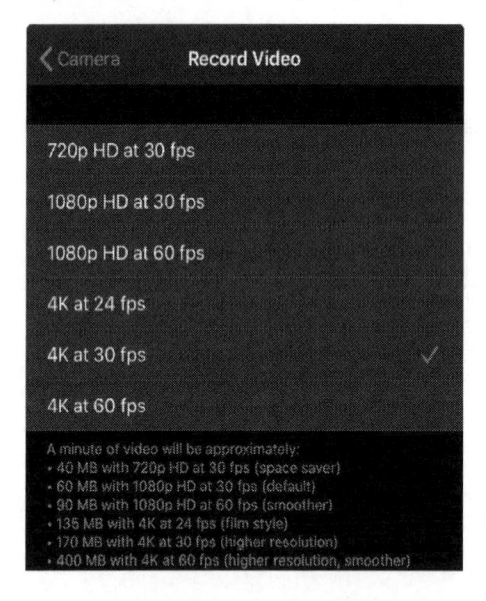

Change Aspect ratio

Launch your device's Camera app and then click on the arrow you will see on the action bar at the top. This arrow will show some buttons in the action bar at the bottom just above the shutter button. Simply click on the 4:3 icon, and select the aspect ratio you want from the list.

Record a QuickTake video

This is the video taken in Photo mode. You will still have the power to adjust the video's record button into a lock position so that you can continue to capture still photos while still recording the QuickTake video.

1. In the Photo mode, quickly touch and then hold the Camera's Shutter button to start taking a QuickTake video.

2. You can get a hands-free recording when you slide the camera's shutter button to the right side and then leave the lock. The shutter button and the record button will be displayed just below the frame.

3. To cease recording, click on the Record button (showing in red).

   Tip: Press and then hold either the volume increase or volume decrease button to initiate taking a QuickTake video in Photo mode. Click on the video thumbnail to see the QuickTake video in your Photos app.

How to shoot a slow motion video on iPhone

Recording a slo-mo video with your phone camera means that you want to slow down the video frame rate so that the time will now look like it is moving at a somewhat slow rate within your video. Videographers normally use effects like this to create a unique video of athletic footage, scenes containing many events that are intense or nature scenes. To have a slo-mo video on your phone camera, follow the steps below;

1. From your device's home screen, launch your device's camera app by clicking on its icon.

2. Enable the slo-mo (toggle on the slo-mo feature) by clicking on slo-mo. The Slow motion video becomes especially very easy on your iPhone 12 with the front camera, and this can be done by clicking on .

3. Tap on the Record button (red button) or deploy either of the volume buttons to begin recording.

4. Click on the camera's shutter button once more to stop the recording.

5. Scroll to your Photos app on your device's Home screen to see the new slow motion video.

How to convert slow motion video into a normal regular video on your device

You can deploy the Photo App on your device to change the slo-mo video clip into a normal regular video. Let us say – for example – that you mistakenly recorded a clip in slow motion and now you want to convert the clip to a normal regular video in real time. To get this done;

1. Launch the Photos app from the Home Screen and go to the slow motion video clip that you plan to speed up.

2. At the lower end of your screen, there will be a slider showing you exactly where the video switches from normal speed to the slow motion speed. You can then drag the small white line situated on the left across the slider till you have totally converted all of the slo-mo areas to regular speed.

3. To eliminate the slow motion effect or to get rid of a small part of the slow motion video, simply move the slider in either direction.

4. Click on "**Done**" when you are satisfied with what you have.

How to convert a normal video to a slow motion video

You can deploy the iMovie app to quickly convert a clip that was recorded normally to a slo-mo video. To do this;

1. Launch your iMovie app from the Home screen and tap on the + icon to initiate a new project with your video.

2. Tap on "**Edit**" to bring the edit screen.

3. When you get to the editing interface, quickly hold down on the part of the clip that you want to slow down. If you plan to slow down the entire video, quickly use your finger to drag across the whole timeline until you see it highlighted/marked in yellow.

4. Click on the speed adjustment icon found at the lower end of the screen. The speed adjustment icon actually resembles a car speedometer.

5. You can proceed to drag your finger across the slider to choose a suitable speed for your video clip. You can actually plan to use one-eight or you can just decide to double the video's current speed.

6. Tap on the **Play** button to watch your video, and when you are satisfied with the result, click on "**Done**" to finish.

Speeding up or slowing down a slo-mo video on your device

The procedures discussed above can be utilized to increase or decrease at which a video is playing using the iMovie app. However, there is no way to get the slo-mo part of the video clip to be slower than your device's frame rate capability. When you are accessing a slo-mo video from your iMovie app, you will be able to see where your video clip begins to slow down right at the lower end of the timeline. If you position your finger on the part that you wish to change, the speed slider will show you how fast or slow the video is playing. If the slo-mo video is already playing at one-eighth speed, slowing it down further will be difficult. However, you can still speed up the video by navigating your finger to the right side on the speed slider.

How to utilize a third-party app to let your video go slower or faster

To get more control over video speed on your device, you can use a third party app like the <u>Slow Fast Motion Video Editor</u>, which you can always get from the App Store.
With the Slow & Fast app, it becomes especially very easy to slow down any part of a video while you speed up other parts. With this app, you can equally trim your video into clips. Then, for each clip, select from −8x to +8x speeds. This provides a rather simple and fast way to make your slo-mo videos go faster or slower than what you can get with the inbuilt Camera app on your device.

The Photo App on iPhone 12 Pro and Pro Max

Navigating the Photo App on iPhone

On iPhone, the Photos app contains a lot of features that help keep proper track of your pictures for easy accessibility. You can swiftly access your photos by year, month, or day, and as well see exactly where the picture was taken. With this, it becomes especially very easy to look for any picture – especially to hold on to important memories.

Moving between years, months and day in the Photos app on iPhone 12 Pro

1. Open the **Photos app** from your Home Screen.
2. Click on the **Library** tab located at the lower end (left) if you are not in the Library tab already.
3. Select by clicking on the exact **timeline view** that you wish to view: **Years**, **Months**, **Days**, or **All Photos**.

   o If you tap on **Years** and then select a particular year, you will be able to view pictures taken in each **months** of that year.

   o If you tap on **Months** and then select a particular month, you will be able to view pictures taken in each **days** of that month.

   o Clicking on a particular picture will bring you all the images from that particular day.
4. You will be able to leave the view mode by clicking on **Days, Months,** or **Years** in the menu bar located above the tab for images.

You can actually swipe from the left bezel of your phone to navigate back to a previous level. But this will not work if you are currently viewing individual pictures. To navigate back, you might have to click on the **back button** found at the top left, because swiping will only navigate to the previous image or to the next image.

Viewing images and location of videos on a map in the Photos app

1. Open the **Photos app** on your device.

2. Ensure that you are within the **Library** tab; this Library tab is the one that will allow you to conveniently move between **Months**, **Years, Days**, and **All Photos.**

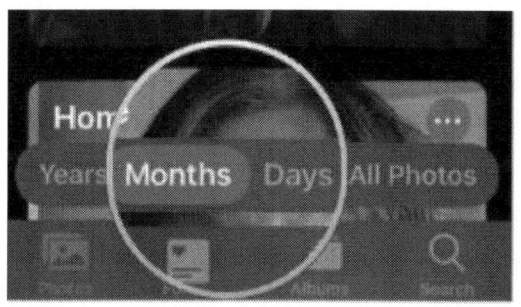

3. If you are within the **Days** or **Months** view, click on the "**more button...**" displayed in the collection thumbnail.

4. Select **Show Map**.

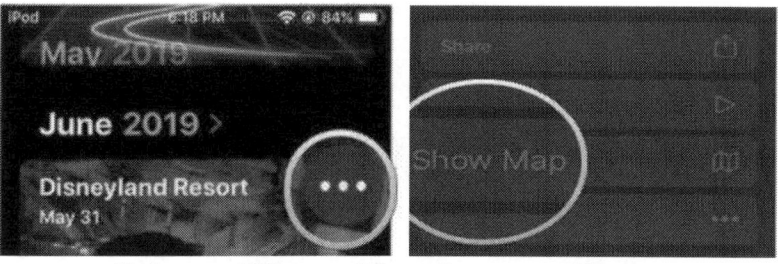

Once you click on "**Show Map,**" you will get a map that will show you the exact location where the image for that month or day was shot. To enjoy this feature, you must have turned on **"Location Services"** for the **Camera app** in **Settings.** When you do this, your device will be able to save geolocation information when snapping pictures.

How to quickly select a month to jump to from the Years view

The previous version of iOS can allow users to access all of the pictures in a particular year at a glance; you only need to start scrolling down on the pictures. But the latest iOS 14 cannot particularly do that, as each of the year is represented with a tile that is able to rotate through a picture from each of the month in a year. Clicking on the tile for the year will quickly move into the month for that particular image in the tile view. The scrubbing gesture can be deployed to quickly jump into a month.

- Open the **Photos app** on your device.

- Ensure that you are in the **Library** tab; the Library tab will allow you to conveniently move between **Years**, **Months**, **Days**, and **All Photos.**

- Click on the **Years** view in **Photos**.

- Locate the **year** that you wish to see.

- Use your finger to horizontally drag across the tile. With this, you will be able to get the month to show under the year, and the thumbnail of the tile will keep changing as you navigate through the months.
- Click on the collection title if you want to jump automatically into the **month** that you left it on while scrubbing.

How to copy your pictures or videos quickly to the clipboard from Moments

1. Open the **Photos** app on your device.

2. Make sure that you are within the **Library** tab.

3. Locate the **pictures** or **videos** that you wish to share from the **All Photos** view.

4. Click on **Select** located in the top right.

5. Mark (tap on them) all the **pictures** or **videos** that you plan to share, or simply **drag** your finger across the columns and rows to swiftly choose a batch.

6. Click on the **Share** icon. The share icon is just a small square that has an arrow pointing upward and it is located on the lower left of the screen.

7. Move down and click on **Copy Photos**.

When you select **copy photos,** the pictures or videos will be sent to the iPhone's clipboard, where you can paste them into a mail, message or document that supports media. If it is only one video or picture that you plan to copy, simply long-press on the picture or video until you get the action menu. Then click on **"Copy."**

Note: Knowing how to navigate and locate your favorite pictures from a pool of picture in the Photos app will let you access your fond memories and share them to friends and families.

How to AirPlay your photos to your TV

If you plan to share a particular video or picture from your iPhone Photos app with persons in a boardroom or your family room, you can use AirPlay to send the video or picture to your Apple TV.

1. From your device's home screen, open the **Photos app.**

2. Click on the **video** or **photo** that you plan to share.

3. Tap on the **share button** found in the lower end (to the left) of the screen.

4. Select **AirPlay**.

5. Click on the **Apple TV** or any **AirPlay-compatible TV** to which you will like to share your picture or video with.

   How to set your iPhone wallpaper using the Photos apps

1. From the Home screen, open the Photos app.

2. Tap the **album** you will like to use pictures from.

3. Open the picture you want to use as wallpaper by tapping on the picture.

4. Click on the **Share button** found in the lower end of the screen.

5. You will see **Use as Wallpaper** among the options on the bottom menu. Simply tap on it to use the image as wallpaper.

6. Click on **Set**.

7. Select the **screen** to which you actually want to set this wallpaper for;

   o Select **Set Lock Screen** if you plan to use the picture as Lock screen wallpaper.

   o Click on **Set Home Screen** if you wish to set the picture as the wallpaper for your device's Home screen.

   o Select the "**Set Both**" option if you plan to use the image as wallpaper for both the Home and the Lock screens.

How to start a slideshow with the Photos app for iPhone

12

There are two methods that you can use to initiate a slideshow in your photos app: you can either select the pictures that you want to use for the slideshow manually, or play the whole album.

How to start a slideshow with selected photos

1. Launch the **Photos** app from the Home screen.

2. Click on a particular **album** or tap on the **Photos** tab.

3. Click on **Select**.

4. Select the **pictures that** you wish to add in your slideshow.

5. Tap on the **share icon found** in the bottom-left side.

6. Choose **Slideshow** from the options. The slideshow will begin immediately.

How to play the whole album as a slideshow

8. From the Home screen, open the Photos app.

9. Tap on the **album** that you will like to play as a slideshow.

10. Tap on the more option "..." button located in the upper right edge of the screen.

11. Move down a little and choose **Slideshow** from the options. The slideshow will begin immediately.

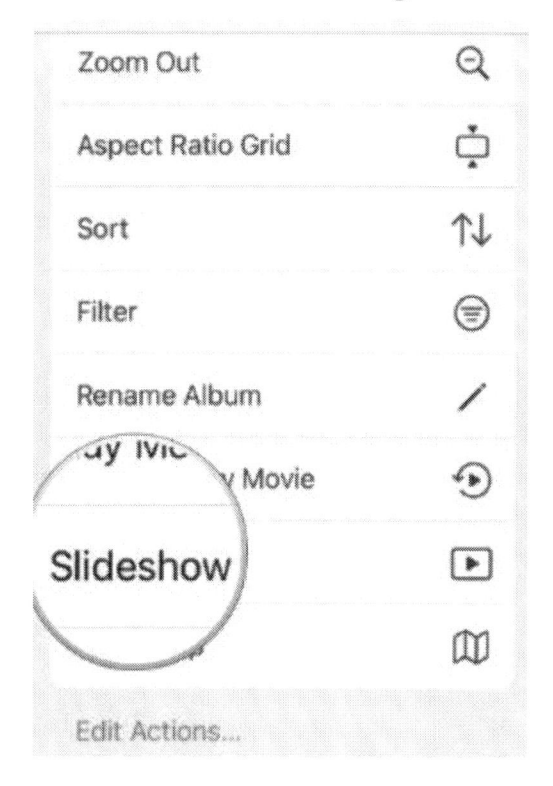

Edit a Live Photo

Edit your live photos by muting the sound and trimming the length of the picture.

1. Open a Live Photo and click on "Edit."

2. Tap 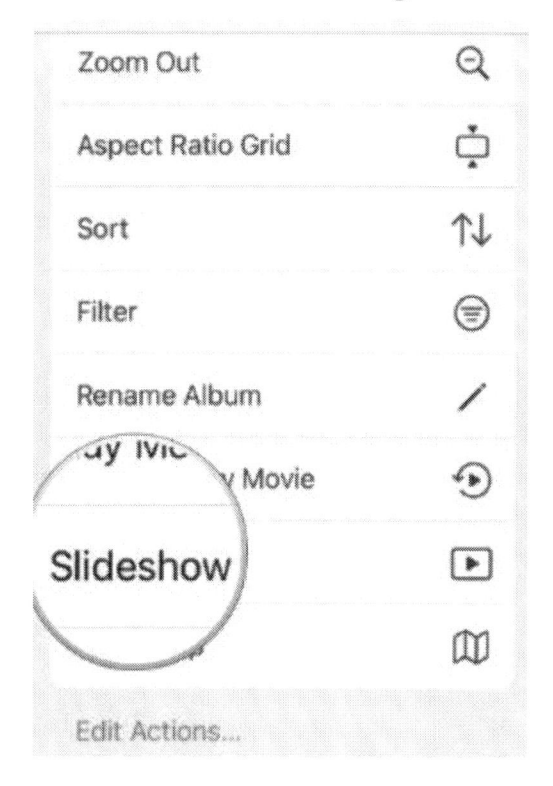, and then carry out any of these settings;

- *Set a key photo:* Simply move the frame that looks like White Square and click on "Make key photo" that will be displayed at the top of the white frame.

- *Trim a Live Photo:* You can adjust the frames for the live image by dragging any of the two ends of the frame viewer.

- *Make a still photo:* Tap on the Live button found at the top part of your screen to disable (turn off) the Live feature. The Live Photo will change to a still photo.

- *Mute a Live Photo:* Click on the mute icon found at the top area of your screen. To un-mute, click on the icon again.

Add effects to a Live Photo

Add effects to your Live Photos to turn them into fun videos.

1. Open a Live Photo.

2. Swipe up on the photo screen to show the effects below your picture, and then select from one of the following: Loop, Bounce and Long exposure.

Play a Live Photo

A Live Photo is any moving image that is able to capture moments just before and right after a picture has been snapped.

1. Open the Live Photo.

2. Tap on the video and hold to play the video.

Play a video

The videos on your phone will auto-play automatically while you scroll and browse through your pictures in the Library tab. Select a video to play the video in full screen without any sound. You can also carry out of these;

- Tap on the player controls just below the video to play, mute, unmute and pause your video. Hide the player control by tapping on the screen.

- You can also double-tap the video screen if you wish to move between the fit-to-screen and the full screen mode.

How to delete and also hide pictures and videos on iPhone

From your Photos app , you will be able to delete images and videos from your phone or hide them from view in the Hidden album. You can also recover pictures you have recently deleted from your device. When you have activated the iCloud Photos, any edited changes will be saved across all your devices that are sharing the same iCloud account with your iPhone.

Delete or hide a picture or a video

In Photos, click on a video or photo, and then carry out either of the following:

- *Delete:* Tap on 🗑 to remove a picture or video from your phone and from all devices sharing the same iCloud Photo logins with your iPhone.

  Any video or image you deleted will be taken to the " Recently Deleted" album for a period of 30 days. During this time, you can decide to recover them back or even delete them permanently from all of your devices.

- *Hide:* Tap on ⬆ and then choose "Hide" from the list of available options.

  Any picture hidden from view will be taken to the Hidden album, and they cannot be accessed anywhere else on your device.

If you decide to disable (turn off) the Hidden album so that it will not appear in Albums, go to the **Settings app** ⚙ on your device, click on "**Photos**," and then turn off Hidden Album.

Adjusting Light and Colors in the Photos App

How to enhance images in Photos on iPhone

1. Launch the **Photos** app on your device.

2. Find the image that you wish to enhance and then **tap** on the image to open it.

3. Tap on the **Edit button** at the upper right corner.

4. You have to ensure that you are actually in the **Lighting** section (it is the dial that has dots surrounding it) and then click on the **Auto-Enhance** button (icon that resembles a magic wand).

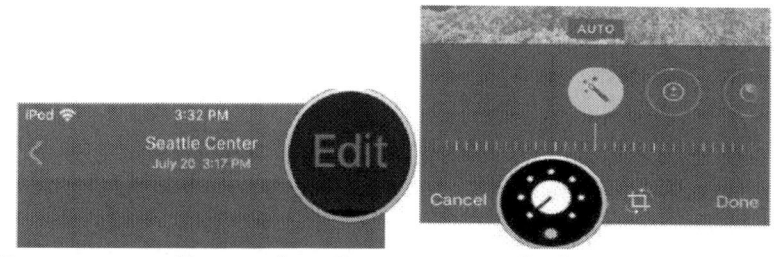

5. You can adjust the intensity of the **auto-enhance** feature by sliding the dial at the bottom end.

6. Select **Done** at the bottom right corner once you are okay with the changes.

How to change the lighting in your photos on iPhone 12

1. Launch the **Photos** app on your device.

2. Open the picture that you want to change lighting for.

3. Tap on **Edit** located at the upper right corner.

4. Tap on the **Lighting** icon at the bottom menu bar.

5. Horizontally swipe through the various lighting aspect categories and select the one that you wish to adjust.

6. For any adjustment that you wish to make, simply slide the **dial** found at the left bottom to right for a stronger or weaker effect.

7. Save all changes by tapping on "**Done**" at the bottom right edge of the screen.

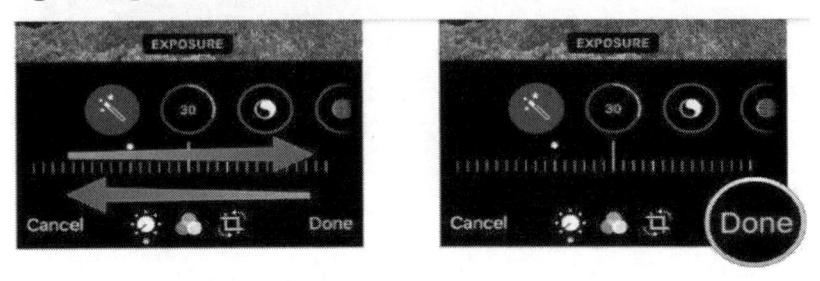

Changing color in your image on iPhone

1. From the Home Screen, tap on the **Photos app** to open.

2. Scroll down to access your photos and click on the picture that you want to change color for.

3. Click on the **Edit** icon.

4. Select the **Color** button located in the middle of the bottom menu bar.

5. **Find a suitable color filter by swiping right and left – the color filter will be applied on your photo immediately.**

6. To carry out further adjustment with the color, click on the **Lighting** icon found at the bottom menu bar.

7. **Swipe** through until you see **Saturation, Warmth, Vibrance**, and **Tint.**

8. The **dial** should be adjusted till you are okay with the final result.

9. Tap on "**Done" located** in the lower end (to the right) of the screen to save changes.

How to convert photos to black and white in the Photos app

The previous version of iOS had an option that can be used to change images to black and white. However, with the new iOS 14, you can no longer use the black and white option directly, but you can always use filters to change a picture to black and white. To do this;

1. Launch the **Photos app** on your device.

2. Locate a **photo** that you want to convert to B&W and **tap** on the picture to open.

3. Tap on the **Edit** icon.

4. Tap on the **Color** icon at the bottom menu bar.

5. Start **swiping** through the filters until you see the three black and white options: **Mono, Noir and Silvertone**. They are applied automatically to your image as you browse them.

6. If you wish to do some adjustments, click on the **Lighting** icon to do some changes to the individual lighting aspects.

7. Select "**Done**" if you want to **save** all changes.

How to revert to the original photo from B&W

Although, converting from a normal picture to a black and white using the Photos app is very easy but you can always revert to the original form of the picture if you are not okay with the result. To do that, follow the steps below;

1. Launch the **Photos app** on your device.

2. Locate a **picture** that you have **edited** with the Photos app.

3. Tap on **Edit** at the upper right edge of the screen.

4. Click on **Revert** from the bottom right edge.

5. **Confirm** if truly you wish to revert the **edited** picture back to the **original** form.

 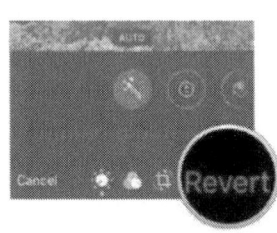 

How to rotate in Photos on iPhone

1. Launch the **Photos app** on your device.

2. Locate the **image** that you want to rotate and then tap on the picture to open it.

3. Click on "**Edit**" at the top right edge of the screen.

4. From the bottom menu, select the **Crop** icon.

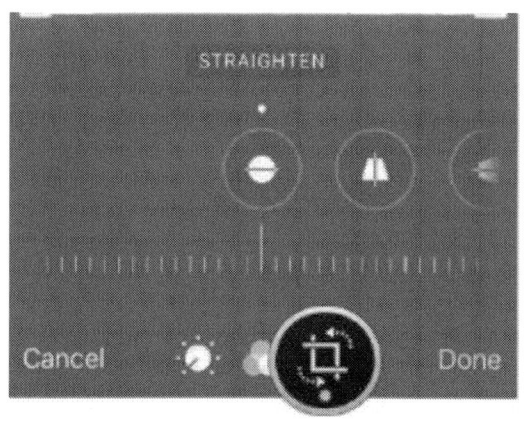

5. Click on the "**Rotate**" button located in the top menu. The **rotate button** is the second button with a curved arrow from the left. Keep tapping on this button till your desired rotation is achieved.

6. Select "**Done**" if you want to save all changes.

How to straighten in Photos on iPhone 12

1. Launch the **Photos app** on your device.
2. Locate the **image** that you want to straighten and then tap on the picture to open it.
3. Select **Edit** from the upper right edge of the screen.
4. Click on the **Crop** icon from the bottom menu.
5. It will enter **Straighten** by default. Quickly use any of your fingers to drag along the **dial** located at the lower end to straighten your picture.

6. If you want to vertically or horizontally straighten out the photo, click on either **Vertical or Horizontal** (found next to the **Straighten** option).

7. Adjust the **slider** and stop adjusting when you are okay with the results.

8. Click on "**Done**" if you want to save all changes.

Sometimes, straightening your image horizontally or vertically is necessary to have a nice perspective of the image. This is true, most times, if you are shooting pictures of buildings and architectural designs, and you need a straight edge on the wall instead of the usual skewed perspective owing to lens errors.

How to change the aspect ratio for crops on iPhone

- Launch the **Photos app** on your device.
- Locate the **image** that you want to crop and then tap on the picture to open it.
- Select **Edit** from the upper right edge of the screen.
- Click on the **Crop** icon from the bottom menu.
- Tap on the **Aspect Ratio** button found at the top (it is the button that has multiple triangles in it next to the "more option..." button).
- Start swiping horizontally to move through the available aspect ratios.
- Select the **ratio** that you plan to use (Freeform is the default).
- Click on "**Done**" if you want to save all changes.

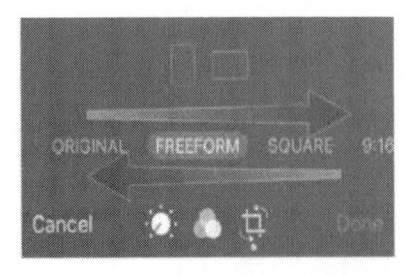 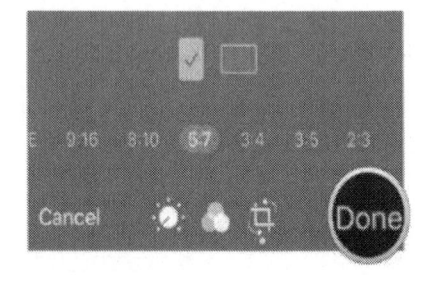

CHAPTER THREE

TECHNIQUE FOR EDITING YOUR PHOTOS USING THE SNAPSEED APP ON iPhone

To download the Snapseed app for free, visit the App Store on your device.

To use the Snapseed app, you will need to open your picture first

Open your photo in the Snapseed Photo Editor

Launch the Snapseed app from the Home screen, and click on open from the top edge when the app opens. This will let you choose a picture from your device's library.

To view your device's photo library and album, choose the "Open from Device" option. Or alternatively swipe through the row of photo thumbnails to choose your recent pictures. Click on the picture that you want to open.

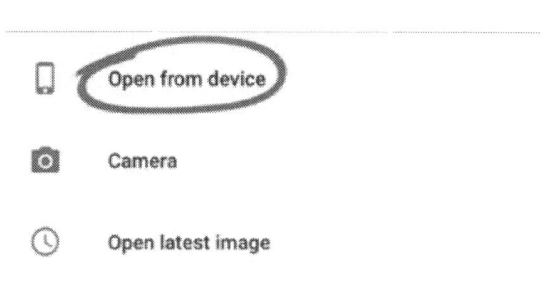

Note: If you select the camera option, you will be able to snap a picture at the moment if that is the option you want to use.

Within the app, you will have access to two Photo editing sections: **Looks** and **Tools**. You will be able to switch between the two modes by clicking on Looks or Tools from the lower end of the screen.

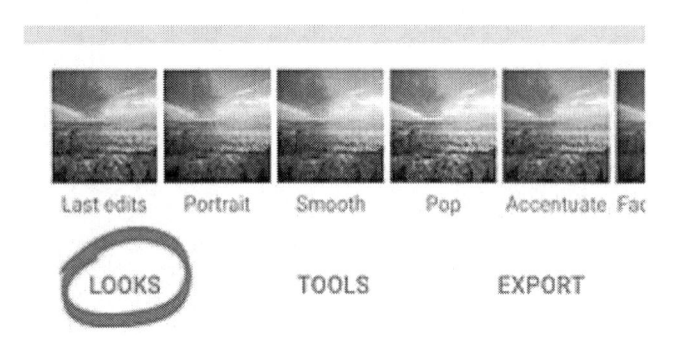

The **Looks** and **Tools** options combined can be used to edit your iPhone pictures to get stunning pictures. The Looks option has other editing tools, which are accessible when you tap on the Look option; and the same thing is applicable with the Tools option.

How to enhance Color, Exposure & Detail in the Snapseed app

Select a picture to open it, and then click on Tools from the lower end. You will have access to the photo editing tools shown below;

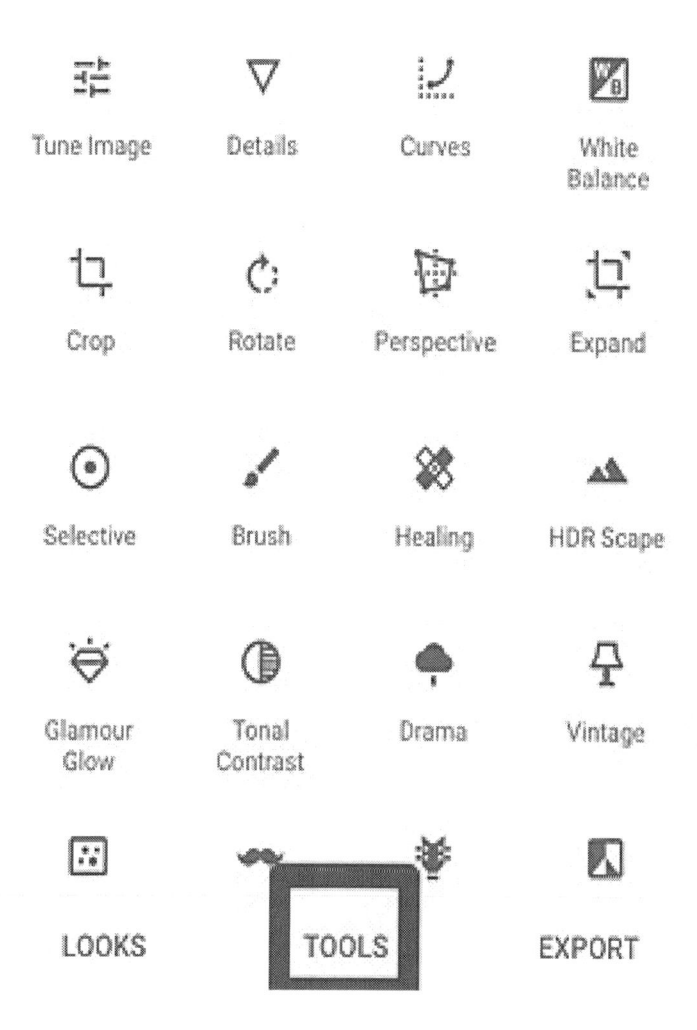

Tap on each tools that you want to use to start editing

Tune Image

With the Tune Image option, you can achieve perfect exposure and color in your image. This tool is used most often in picture editing using the Snapseed app because it can improve the look of any picture to a good extent. The example below depicts how you can convert your dull and dark image into one with amazing color.

To use the Tune image, simply select the Tune Image option from the Tools section in the Snapseed app. Launch the Tune Image menu by swiping up or swiping down on the photo. Choose the tool that you plan to use; choose between Contrast, Brightness etc.

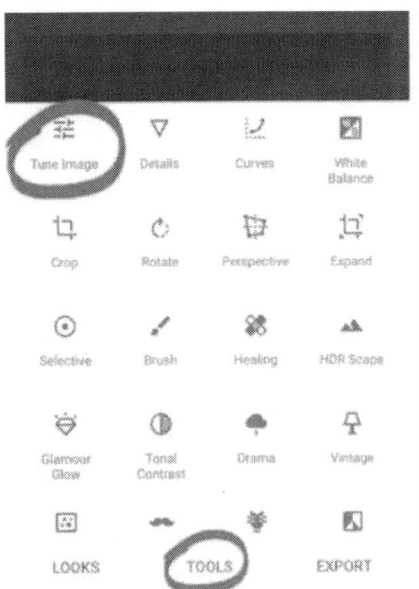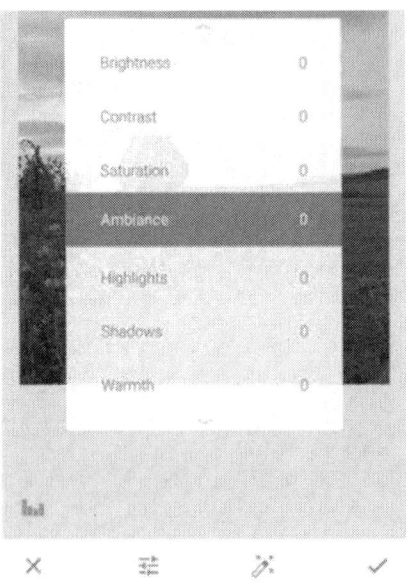

You can take a look at what happens to the exposure and color of your picture when you select the Tune Image tools;

Brightness: When you select this, the whole image can either darken or brighten depending on the initial state of the picture.

Contrast: Increase or reduce the difference between the dark areas and the bright areas of your picture.

Saturation: Choose to let the colors be more vibrant or less vibrant.

Ambiance: Modify the color saturation and color contrast also.

Shadows: Darken or brighten only the dark areas (shadow) in your image.

Highlights: Darken or brighten only the bright areas (highlights) in your image.

Warmth: Use a cool blue color cast or a warm orange color cast in your picture.

When you have selected your desired picture, swipe to the left or right over the image to carry out necessary adjustment. The setting value will be indicated at the top area of your screen.

The above example indicates how the color vibrancy can be adjusted using the Saturation option.

The image below is showing you how the Warmth option can be used to cool down or warm up your image. The image at the left was set to warmth (+25) while the one on the right was set to warmth (-25).

You can use any other editing tool from the Tune Image menu by swiping up or down, select another tool and then adjust settings by swiping left or right.

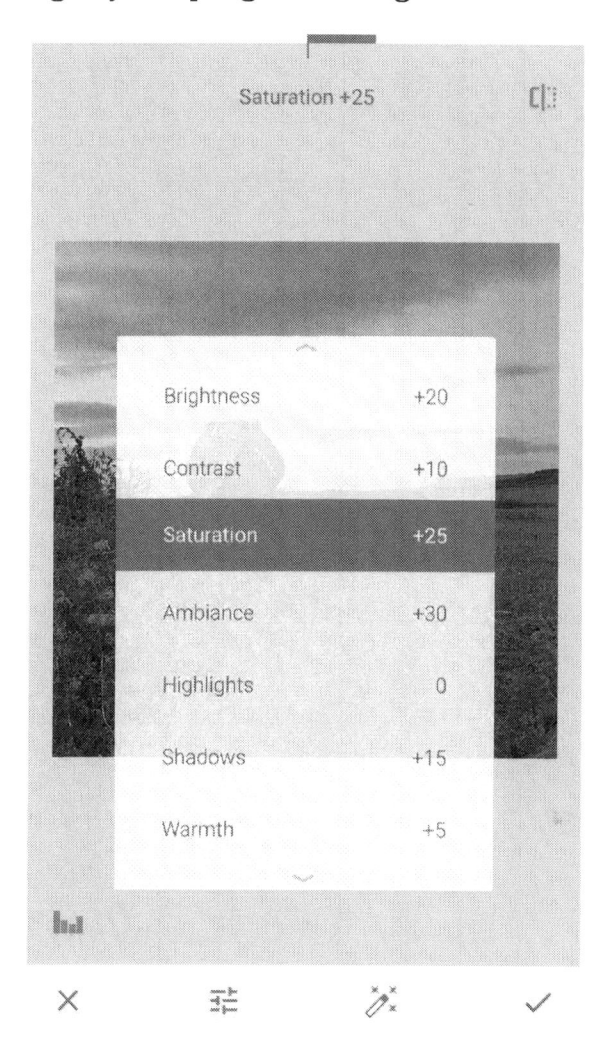

The "before and after" versions of the image can be compared by tapping on the icon at the top right side. Tap and then hold on the icon to view the original picture, and then return to the original picture by releasing your fingers.

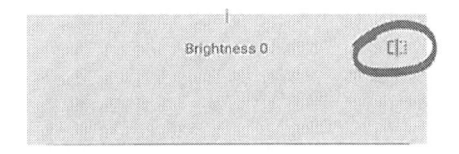

Click on the checkmark once you are satisfied with the result.

White Balance

Use the White Balance tool to enhance your image by using different tints of colors.

With the color tints, you will be able to correct unnecessary color casts, enhance colors or change the mood of the picture. From the Tools section of the Snapseed app, select White Balance. Open the White balance tools by swiping up or down, and then choose a setting from the menu.

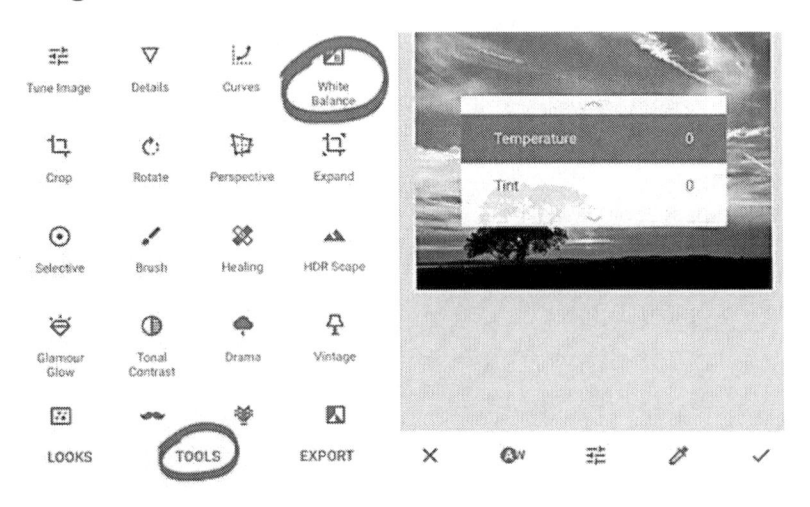

Two white balance tools exist in the Snapseed app;

Temperature: Achieve cooler (blue) color or warmer (orange) color in your image.

Tint: Use a pink or green color tint in your photo.

When you have selected a tool, quickly swipe to the left or to the right to adjust settings.

If you look at the picture above, the Temperature option has been used to warm up or cool down colors. Warming up colors is okay for enhancing sunset pictures.

You can also use the Temperature option to correct cool or warm color cast in your picture. In the image below, the snow looks blue. You can increase the Temperature to warm the color and let the snow appear white instead. For the image on the left, the temperature has been set to zero, while the one on the right was set to (+30).

Another example below shows how the Tint setting has been deployed to bring out the true green color in the leaf. Compare the picture on the right with that on the left and you will see the differences. In the picture on the left, the Tint has been set to zero, while Tint was set to (-40) for the one on the right.

Click on the checkmark at the lower left to apply the settings when you are done.

Details

Improve the fine details and the texture in your picture with the Details tool. This tool is best deployed on images that have amazing textures like wood grain, paint, peeling etc. From the Snapseed Tools, select Details. Choose the settings that you wish to change by swiping up or swiping down.

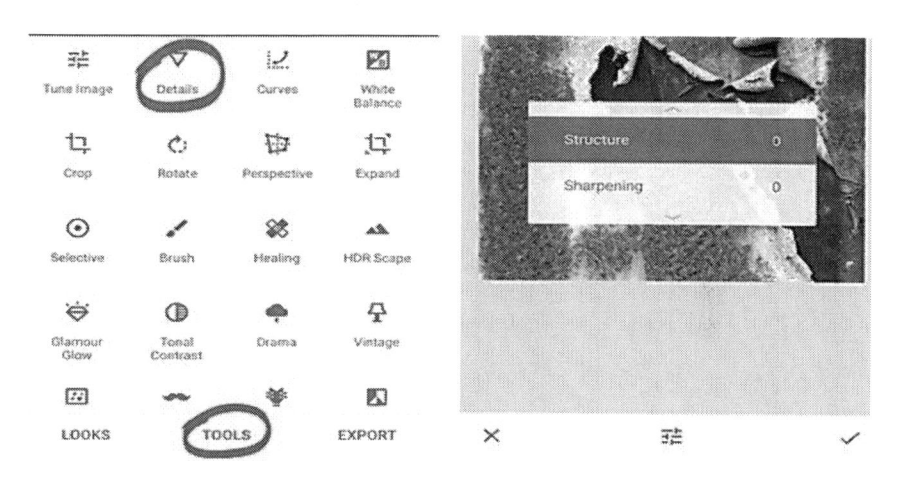

The **Structure** option is most times preferable because the **Sharpening** option decreases image quality by adding grain in the picture. Swipe across your screen to modify the settings, just like the other tools you have been using in the Snapseed app. When you are editing with the Details tool, you can zoom in to view the adjustment you are making very clearly. Zoom in by placing two fingers on your screen, and then drag apart your fingers.

The Details tool, although capable of giving your picture that wow factor you desire, but doesn't necessarily improve the entire image. For instance, bringing out the texture in a portrait image can make the skin appear bad and make the picture look a little bit older.

Techniques for Retouching Your Photos

It has been shown, at least times without numbers, that the iPhone 12 camera can be used to take interesting videos and photos – videos and photos that are of almost the same quality as those obtainable with a digital camera. But due to the nature of taking pictures with a mobile camera; the spontaneous nature and the spur-of-the-moment nature of shooting and recording videos, there might be errors during this process. When you take out your device to record a scene or event, it is likely you have issues pertaining to composition, light and color. These problems can be fixed with the Photo apps on your iPhone, and you can even have access to more advanced capabilities by downloading and installing a wide range of photo and video editing apps from the App store.

In this section, you will see how to fix some of the basic photography glitches that might arise while using an iPhone camera to take pictures.

Fix a Photo's Colors

Tools required: Photos app and Adobe Lightroom app
One main setback usually experienced while taking pictures with your device's camera is poor lighting, which can really result in bad color conditions for your picture. This is unavoidable while taking pictures with an iPhone

camera since you might be required to attach some external light instruments and sources to fix the glitches. In essence, when you are using an iPhone camera, you may need to edit the color of the picture (after the picture must have been taken) as this is one of the easiest troubleshooting techniques to achieve the right color for your image. Basic color editing can be in the form of changing a photo property, such as contrast, saturation, temperature, hue etc. There are many photo-editing applications that can allow you to edit the color settings for your photo. In this section, you will learn how you can modify your image colors with the Photos app on your device and with the Adobe Lightroom app that you can actually download from the App Store for free.

Modify Colors Using the Photos App

With the latest updates of the iOS (recent iOS 14 update), the Photos app now has some basic photo adjustment tools that can let you modify your pictures without necessarily downloading and installing any third party editing software on your device. The tools in the Photo App can provide some basic editing capabilities if you don't need much editing for the picture. Proceed with the steps below to edit your pictures with the Photo App;
1. Launch the Photo App from your device's home screen and select the picture that you wish to edit to display it in full size.
2. Tap on the adjustment icon found at the bottom of the screen (a three horizontal slider line beside the trash icon at the bottom of the Photo App's screen). This enables you to use photo-editing mode, as shown below;

3. Tap on the editing icon, as depicted below, to access the edit options for Light, Color, and B&W. The Light options will enable you to modify the light in your picture; the Color will allow you to modify the properties of the color, while the B&W will allow you to convert your photo to black and white.

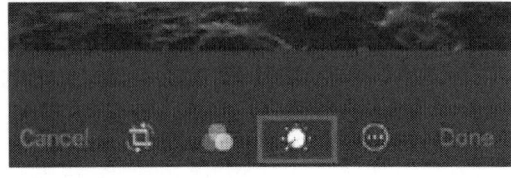

4. Tap on the Color heading to expand it and use more functionality.
5. You will be able to increase the colors saturation by dragging the Saturation slider to the left.
6. Click on the icon located at the top right of the color heading, and select Contrast in the Color section.
7. Drag the slider to the left hand side to increase the color of the contrast.

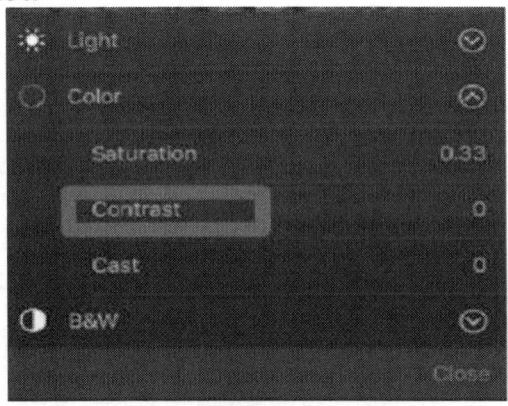

Remove Photo Casting

The light situation in the area where you are snapping your photos will usually affect the color of the final photos thereby creating a color casting effect. This is possible most times when you are shooting your photos in some light conditions, like taking pictures in fluorescent light, on cloudy days or in warm light. The color casting effect can disorganize the clear and natural color in an image. In addition, you may even decide to use a cast color to achieve some special effects in your picture, for instance, adding a warmer tone to the photo. Use the steps below to remove or add a color cast in your photo;

- Choose the adjustment icon (looks like three horizontal slider lines beside the trash icon at the bottom of the Photo App's screen) in the Photo App, and click on the Color section.
- Click on the icon found at the top right section of the color slider.
- Click on "Cast" in the Color section heading as displayed below;

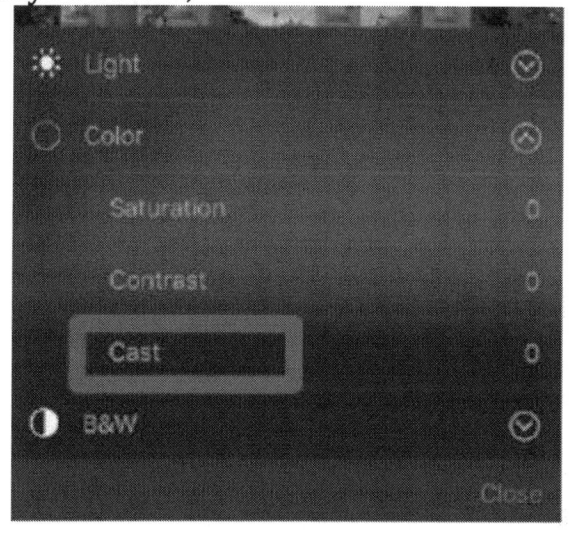

4. Achieve a warmer effect when you drag the slider to the left and a cooler effect can also be obtained when you drag the slider to the left as shown in the screenshot below;

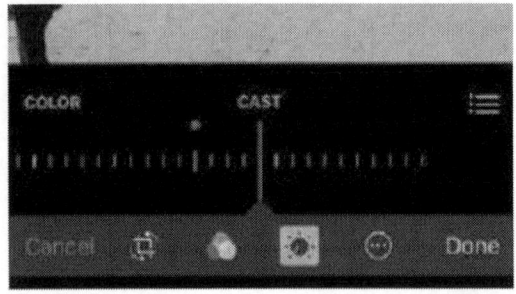

5. Click on **"Done"** when you are okay with the result.

Make a Color Adjustment Using Adobe Lightroom

Adobe Lightroom is one of the frequently used and most powerful applications for photo editing and photo management on iOS devices. It has advanced capabilities that you can deploy for modifying your images and the details of your photo. It is most commonly used by professional photographers and persons who specialize in photo editing. A free version of the Adobe Lightroom app is available for download on the App Store. A free Adobe ID is required to use this application and you can create one on the app. The free Adobe Lightroom version will allow users to snap, arrange, and share photos from their mobile phones. In fact, with an Adobe Creative Cloud membership, you will be able to use the extended features of the Adobe Lightroom application, and you can even save the source format allowing you to work with both the desktop and mobile version.

With a free account on the Lightroom app, you will only have access to basic adjustment tools. To use the Adobe Lightroom app, you will need to download and install the app from the App Store. Once you have successfully downloaded the app, open the app from your device's Home Screen to start editing.

How to remove unnecessary spots and any unwanted objects in your picture

The **Healing Brush** tools shown above can be deployed to eliminate any unwanted objects, spots, people, power lines, or some other side distractions in your image.

1. From the **Edit** panel in the lightroom app (Loupe view), tap on the **Healing** icon found at the lower end of the screen as indicated above.
2. You can then go ahead and use any of these **Healing Brush** tools:
   o **Heal**: Tapping on this will bring the texture from the source area and then matches the texture with the tone and color of the target region in the photo.
   o **Clone**: Apply the same pixels from the source region in your photo to the target area.

Both the **Heal** and **Clone** tools utilize the texture obtained from the source area and then apply it to the target area. The **Heal** tool, however, takes the tones and colors that surround the target area and then blend all of these together, while the **Clone** will produce the pixels from the source section to the target area.

A. =Heal **B.** = Clone **C.** = Size **D.** = Feather **E.** = Opacity **F.** = Delete **G.** = Target area **H.** = Source area **I.** = hide screen controls to view the photo edits

By choosing the **Heal (A)** or **Clone (B)** tool above, you will be able to brush over the item in your image that you wish to retouch or delete. Once you have been able to brush over the item in the picture, you will have two marquee areas. One of the white marquee areas over the painted object will be showing the target area, while the second white marquee area that has an arrow pointing at the target area will indicate the source area.

You can then go ahead to modify the size, feather, or opacity of the selected **Healing** tool as appropriate.

✓ **Size.** Indicates the brush's tip diameter in pixels.

✓ **Feather**. Carry out a simple soft-edged transition between the brushed area and the surrounding pixels in the target area.

✓ **Opacity**. The opacity of any adjustment you have applied to the target area can be adjusted with this tool. Touch the left controls, and then adjust the value by dragging up or down on the screen.

If you want to move and also position the source or target area on your image, quickly drag the blue pin at the middle of that area. Click on the (⬀) icon located at the upper-right corner to see the photo edits on a fullscreen. This will hide the screen controls at the bottom and the white source/target areas.

Crop photos

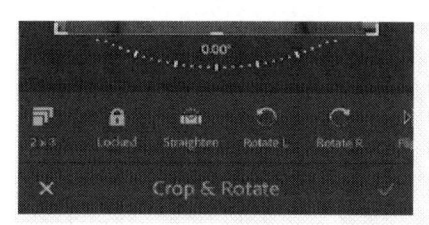

1. From the **Edit** panel in the Loupe view, choose the "**Crop**" icon found at the lower end of the screen.
2. You will see the available Cropping options as tiles just along the bottom end of the screen.
3. You can explore all of the tiles by swiping to the right or left. Click on a tile to use the corresponding option.

Adjust the tonal scale of a photo

You can deploy the tone controls in the **Light** menu to quickly adjust the tonal level of your photos. To carry this out, proceed with the steps below;

1. From the Edit panel, tap on Light icon from the bottom end of the screen to bring the tone controls.
2. Though not important, you can choose Auto to adjust the entire tonal scale.
3. Adjust the tone control sliders:

Note: You can use two fingers to gently tap on the picture to bring the histogram. Start looking at the histogram while you work to adjust the image's tone control to visualize what is actually happening.

From the tone control slider, the following settings can be adjusted;

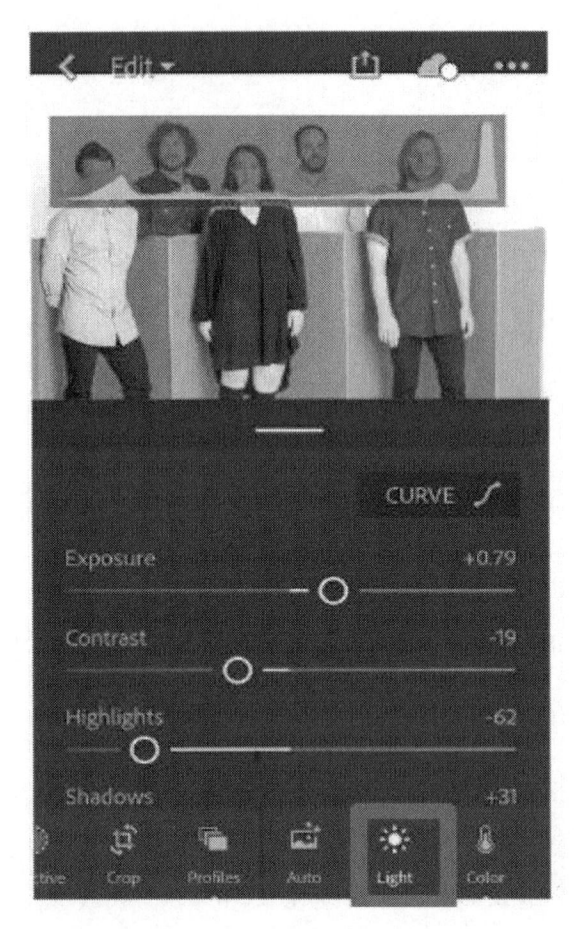

Exposure

Adjust the whole brightness for your picture. Start adjusting the brightness slider until you are satisfied with the appearance of the picture and stop when you have the intended image brightness.

Contrast

Use Contrast to increase or decrease the image contrast, majorly affecting the midtones. When you increase the contrast, you will get a darker middle-to-dark image, and the middle-to-light image parts will actually become lighter. As you reduce the contrast, the tones of the picture will be affected inversely.

Highlights

This can be used to modify bright areas in your photo. You can drag the white circle (see the above screenshot) to the left hand side to darken highlights and also to recover "blown out" highlight details. You can as well brighten the highlights by just dragging the white circle across the highlights slider.

Shadows

The shadow option can be utilized to provide some adjustment to dark areas in your pictures. You can darken the shadow by dragging the white circle on the shadow slider to the left. The white circle can also be dragged to the right hand side to brighten shadow areas and to recover shadow details.

Whites

You can use this to modify white clippings in your pictures. Increase the white clippings by dragging the white circle to the right, and you can reduce the white clippings by dragging the white circle to the left.

Blacks

You can use this to modify black clippings in your pictures. Increase the black clippings by dragging the white circle to the left, and you can reduce the black clippings by dragging the white circle to the right.

Working with Photography projects

Most DSLR cameras, today, allow you to use different photography techniques to improve the quality of your picture. You can quickly use effects like the shallow depth of field and double exposure in your DSLR camera without using computer software like the Photoshop or Adobe Lightroom. While it is difficult for an iPhone to have the same tools you have in your DSLR camera, the good news is that you can use many third party apps to achieve the same effect you have with your DSLR camera.

Creating Advanced HDR Photos

Tools used: Snapseed app

The high dynamic range (HDR) effect when applied to an image will raise the contrast between the shadows and light in your pictures by giving you a wider range of luminance when compared to normal images. This allows you to have images with dramatic effects. While it is possible to create a good HDR image with the Camera app of your device, you cannot really apply some advanced effect from your camera app in your HDR image. Hence, there is a need to use a third party app that is capable of controlling the style or density of the HDR effect you have applied.

HDR photos are normally obtained by merging many pictures having different exposure levels to get a wider luminance in the scene. Meanwhile, to have a better result with the HDR effect, the image must include colorful elements with high-contrast shadows and lights.

Step 1: Prepare the Photo
Before you apply the HDR effect, you will need to ensure that the colors of the photo are contrasted and vibrant enough to get improved final results. Follow the steps below to increase the contrast, brightness, and saturation in your shot;
1. Launch the Snapseed app. Select Open and then open a picture on your device.
2. Tap on the Tools button located at the bottom right, and select Tune Image.
Tap on the Adjustment icon, and select Brightness. Drag the slider located at the top to +20 in order to increase the brightness of the picture.
3. Tap on the Adjustment icon, select Contrast, and assign it a value of +50.

4. Tap on the Adjustment icon, select Saturation, and then increase the saturation to around +30. Also, you will need to set Ambiance to +30

5. Tap on the Apply icon located at the bottom right to deploy the effect.
6. Tap on the Tools button on the bottom right, and select White Balance. With this, you will be able to apply warmer color to the picture.

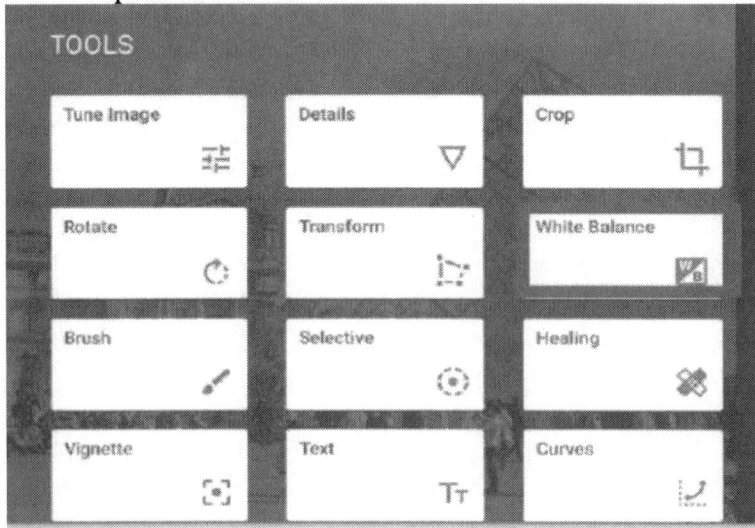

7. Tap on the Adjustment icon, and select Temperature. Set the temperature to a warmer temperature, for instance, +15. Click on the Apply icon

Step 2: Apply the HDR Effect

The ability to select the HDR effect that is suitable for the picture you are working on will go a long way toward determining the quality of your picture. To do this, follow the steps below;

1. Tap on the Tools icon, and select HDR Scope from Filters.

2. Select the Strong mode, and then increase the strength to +50

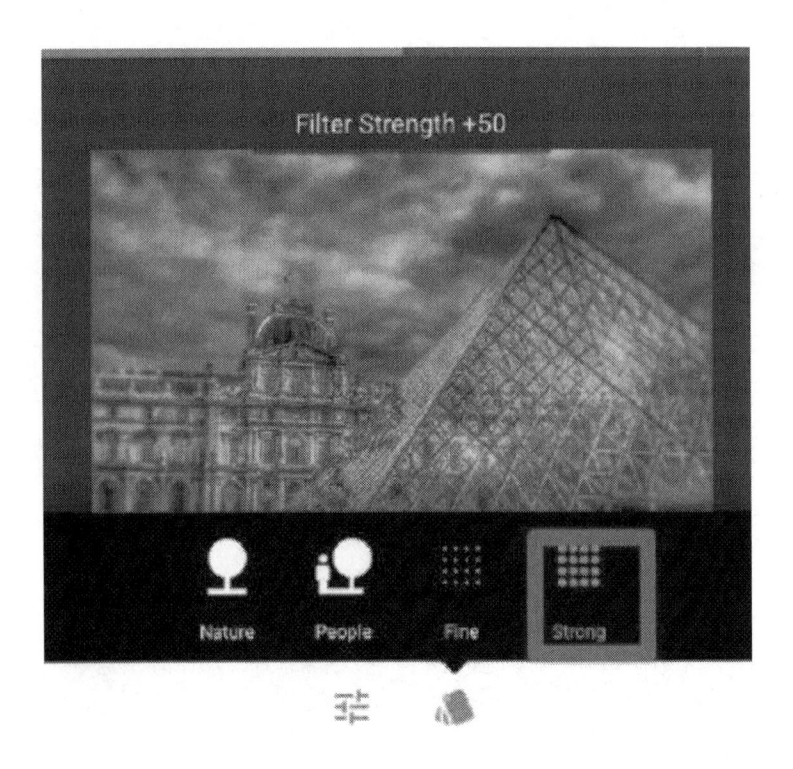

3. Tap on the Adjustment icon, select Brightness, and then set the brightness to be a bit brighter by dragging the Brightness slider to +10.

4. Click on the Apply icon.

Step 3: Apply the Vignette Effect

Add a vignette effect and have dark edges for the picture in order to add more focus to the center of the image and the pyramid appearing in the background. To do this;

1. From the Tools list, select Vignette. Make dark edges in the picture by dragging the slider to the left.

2. Tap on the Apply icon.

Step 4: Add the Blur Effect

After applying the vignette effect, a blur effect will be added to increase the focus at the center of the image. To add blur, simply follow the steps below;

1. From the Filters list, select Lens Blur. You can increase the strength of the blur to around +30.

2. With your two fingers, adjust the shape and size of the blur by dragging. Let the blur be only on the edges so that you will have the center point of the image in focus.

3. Tap on the Apply icon

4. Click on Save to save all changes. Then you can save a copy to preserve your original image untouched. When the image is saved, you can always open the picture again to edit the modifications. If you select Export, the image will be exported, yet all the changes can no longer be edited.

Creating a shallow depth of field

Tools used: Tadaa

The word *depth of field* talks about the focus range in your image. If the focused area of the image is restricted to near objects while the elements that are far look blurry, then you can say that you have a shallow depth of field. However, if the focused area of the shot contains a wide area of the picture, then you can say you have a deep depth of field. The camera aperture (f-stop) is the one controlling the depth of field in your DSLR camera, and this has something to do with the diameter of the lens, denoted as a fraction (1/F). If the lens opens large, this number will be small, and this results in a shallow depth of field.

On your iPhone, you have a limited aperture control since you are only able to tap on the item you want to focus on. To have more control over the depth of field, you can use a third party app.

Step 1: Protect the Main Objects

Before you apply the blur effect, you will need to protect the main objects in the picture to keep them sharp. The mask effect can be deployed as follows;

1. Open a picture in your Tadaa app, and select the Blur icon located at the bottom toolbar.

2. Click on the Edit Mask icon located at the top.
3. Zoom in the objects and begin to paint a mask over the object.

Tap on the Apply icon located at the top right.
Step 2: Apply the Blur Effect

Tap on the Blur icon. You may have to drag the toolbar to the right or to the left to show the other part of the tools

Use the Circular option as the blur type.
3. Drag with your fingers to resize and position the protected area above the near objects in the photo.

4. Increase the effect of the blur on the far item by dragging the blur slider. The level of blur is dependent on visual distance between the far object and the near objects.

5. Lower the Range value to begin the blur effect from a proper distance from the in-focus objects.

Tap on the Apply icon to apply changes.

Step 3: Apply the Vignette Effect
Now, you will darken the edges of the photo to focus on the main objects as follows:
1. Click on the Vignette icon located at the bottom toolbar.
2. Drag the Intensity slider to your left side to have a proper frame around your photo.
3. Adjust the Range slider to the middle point

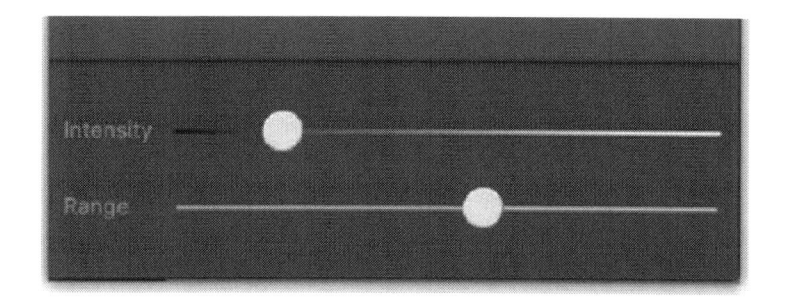

4. Click on the Apply icon found at the top right side.

Step 4: Apply a Final Filter

Once you have successfully added the vignette effect, you can get a cool and dramatic mono color effect by adding a filter to the picture. To do this, follow the steps below;

1. Click on the Filter icon, and then select Charleston filter.
2. You can lessen the effect of the filter by dragging the filter toward midpoint.

3. Click on the App icon, and then select the Share icon if you wish to save the image to your Camera Roll.

CHAPTER FOUR

THE CAMERA LENS: THE BASICS

Focal length basics

If you have ever handled a lens, you must have seen the "mm" boldly written on it and you can only wonder what this means on the lens. The letters "mm" are usually accompanied by a number or some set of numbers next to it. The number written with the "mm" is the focal length of the lens you are holding. When you take a look at the lens below, the focal length is 8-15mm. What this means for the lens is that you will be able to change the focal length within the 8 to 15mm range.

This change is often called variable focal length or zoom.

You will be able to zoom in from 8mm and zoom out from 15mm. This is the actual range of the variable focal length for the lens above. For the lens, you won't be able to go below 8mm and you cannot also go beyond 8mm.

For camera lenses, the focal length will allow you to take shots and capture footages that have different visual characteristics, but these properties will be able to influence the effect the picture or the footage will have on its viewer.

Photographers and professional videographers choose video or photo camera lenses for various reasons, hence getting to understand how the focal length of the camera lens affects the overall look of the picture might be one reason why you can consider going for a particular lens over the other.

Focal length is the actual distance between the camera sensor and the optical center of your lens when focused at infinity. The exact point where the light ray converges in your lens is referred to as the optical center of the lens. The focal length is what actually defines the field of view and the magnification for a particular lens, and its value is normally measured in millimeters. A zoom lens has variable focal length while prime lenses usually have set focal length, and the visual properties are always affected by any change in magnification.

Why, then, is the focal length important in choosing a lens?

- It is the focal length that usually determines what part of the image will be in focus for proper recognition. Essentially, it helps to properly isolate

items in the background, foreground or the middle ground.
- It helps to produce visual contexts for your image
- Influences the visual appeal of your shots.

Let us put things in proper way; if you have a longer lens, a narrow image closer to your subject is produced. However, for a wider lens, more of the area is captured from the very back.

A lens that has a focal length around 35mm to 55mm will be categorized under a **standard** focal length. The reason is that the human eye operates at about the same field of view. The choice of focal length actually matters for both photo lenses and video lenses; this is because a video or a picture that was shot with a 50mm lens will have a somewhat different look and appeal compared to one that was shot with a lens of 20mm.

The Camera lens aperture

Aperture is used to talk about the light intensity of a particular image or group of images. Lens aperture is the one that controls light entering through your camera lens to the film or the image sensor. The aperture is measured in F-Stops or T-Stops.

The T-Stops are normally used when comparing the best camera lenses with a digital photography lens. Camera lenses having higher stop ratings with wider openings will enable more light, which allow photographers to shoot in dark places. The aperture ratio of your camera lens will usually be expressed by the lens ratio (this is focal length/max aperture).

The shutter speed

The amount of time the camera shutter is opened is referred to as shutter speed. The shutter speed is normally measured in fractions of a second, most especially when filming.

When you are doing still photography, you have a much less rigid shutter speed than what is applicable in cinema. This is because still photography is seen as a single frame.

What is depth of field?

Depth of field talks about how much of the image is in proper focus, most specifically the area of focus that is acceptable.

A blurred background will arise with a depth of field that is shallow, while a large depth of field will usually make the whole picture come into focus so that you will be able to visualize everything surrounding the image.

Type of Camera Lens

The quality of your picture is determined by many factors; one of which is the quality of lens you attach to your device. Sharp image having many details and contrast in the background is the hallmark of a camera lens that can be considered good, while a poor camera lens will most times make your picture look blurry and dull. When you want to buy a camera lens for your device, it is pertinent that you take the kind of object that you want to be shooting into adequate consideration. You will also need to consider – of course – your budget and the lightning condition that you want to be shooting the image with. One thing you should know here is that no single lens is perfect for every condition; hence, you will need to weigh your options and know where you want to compromise.

Wide-angle iPhone Camera Lens: With the Wide angle lenses, you can always get more than what you can get with the normal lens that came with your iPhone. The Wide-angle lens is good, especially for landscape pictures, and can even provide substitute for taking panorama picture instead of using the iPhone camera lens.

Telephoto iPhone Camera Lens: With a lens like this, you can always bring the subject in closer range to the camera. The bigger the magnification, the larger the object distance will be in your image. With a Telephoto lens, you will be able to have a better flattering portrait image with a more natural facial feature.

Fisheye iPhone Camera Lens: Fisheye lenses produce spherical images that look dramatic, with distorted lines and darkened corners. The distortion produced in these photos contributes to their appeal. One cons of this lens is the fact you really have to move closer to the object to take a nice shot. The Fisheye lenses are good for taking creative shots.

Macro iPhone Camera Lens: With lenses like these, you can take pictures of subjects that are not visible to the ordinary eye. To have detailed and in-focused shots, you will have to move closer to the subjects and hold your phone in still position.

What is a smartphone lens?

With more and more professional photographers and videographers growing by the day, it becomes especially important to consider other alternatives to taking cinematic shots at a glance without the need to carry heavy DSLR cameras from one location to another. Hence, there is increased preference for iPhone photography since iPhone cameras are able to take clear and amazing shots; most especially when coupled with add-on lenses. One of the best smartphone lenses is the Moment lens.

If you want to attach a moment lens to your iPhone camera, you will need to use the Moment Brand case to be able to attach the lens conveniently to your iPhone.

Olloclip ElitePack for latest iPhone

This lens is compatible with latest iPhones (iPhone SE and later). There are 3 lenses in the pack. The lens type is that of **telephoto, fisheye** and **macro lens.**

To get the best out of your iPhone 12, you should invest in the Olloclip lens for iPhone. Inside the kit, you will get a 2x telephoto lens for more coverage, together with one other lens consisting of two in one : a fisheye (180 – degree) and a 15x Macro. With the complete Olloclip set, you will have real versatility which actually compensate for the high price (129 USD) of this lens.

Moment Anamorphic lens

The lens type is Anamorphic lens having only one lens. If you are shooting videos with your iPhone 12, this lens is a go to lens for you to have that perfect cinematic effect in your video. The image produced with this lens is of high quality.

Xenvo Pro lens kit

The lens types in this kit are of the **Macro** and **Wide-angle** lens. The number of lenses is two (2). The **Macro** lens enables you to take impressive close-up photographs. It also comes with an OLED light to aid close up photography. It goes for 39.99 USD on Amazon.

Moment Wide-angle lens

The lens type here is that of the Wide-angle lens and it has only one lens. You will need a Moment M series case (which is sold separately) in order to fit this lens perfectly on your iPhone 12. The Moment lens produces high quality shots and it is a good investment if high quality photography is your thing. It goes for 119.99 USD on Amazon. The Moment Wide-angle lens and the Moment M series case are shown below;

Moment FishEye Lens

The Moment FishEye lens works just about the same way as the anamorphic lens and the Moment wide lenses, but it produces a comparatively wider view. In actual fact, it produces a 50° wider field of view when compared with the ultrawide lens. It provides the coverage of that of a traditional fisheye lens (14mm), while filling your frame with a distorted picture to have dramatic effects. This lens is especially okay for shooting pictures in tight places. It has the ability to shoot in ultra-wide 170°, which will be good for those who need to shoot architectural designs or interiors – or people who want to shoot a cartoon-like image.

Apexel HD 1.33X Anamorphic Lens

This one is more affordable if anamorphic lens category is your thing. It gives an amazing cinematic shot that has little distortion.

Apexel 36x telephoto lens

If you want an amazing telephoto shooting, then this lens will provide you the best reach. It can transform your device into a better long-distance shooting gadget. It also has a small tripod that can be used to stabilize your shots.

It also comes with a remote shutter that can be used to take pictures at a very long distance. It is not expensive.

About the Author

Konrad Christopher is a video software expert with several years of experience in video-graphy and software development. He is consistent following the latest development in the tech and software industries and has an eye for highend video equipment and software. He loves solving problems and he is enthusiastic about the software market.

Konrad holds a Bachelor's and MSc degree in software engineering from Cornell University, Ithaca. He lives in New York, USA. He is happily married with a kid.